SNOWBALL EARTH

SNOWBALL EARTH

THE STORY OF A MAVERICK SCIENTIST AND HIS
THEORY OF THE GLOBAL CATASTROPHE THAT
SPAWNED LIFE AS WE KNOW IT

GABRIELLE WALKER

BLOOMSBURY

First published in Great Britain 2003
This paperback edition published 2004

Copyright © 2003 by Gabrielle Walker

Bloomsbury Publishing Plc, 38 Soho Square, London W1D 3HB
Published in Association with Crown Publishers, New York

A CIP catalogue record for this book is available from
the British Library

ISBN 0 7475 6850 2

10 9 8 7 6 5 4 3 2 1

All papers used by Bloomsbury Publishing are natural, recyclable
products made from wood grown in well-managed forests.
The manufacturing processes conform to the
environmental regulations of the country of origin.

Printed in Great Britain by Clays Ltd, St Ives plc

www.snowballearth.com

FOR

ROSA, HELEN AND DAMIAN

CONTENTS

CONTENTS

ACKNOWLEDGMENTS

For the past two years or so, I have been a Snowball Earth groupie. Wherever the story was told or challenged, at conferences, field trips, lectures and campsites around the world, I appeared with my notebook and questions. Some of the researchers I spoke to are mentioned by name in the book, while others are not. All gave generously of their time and knowledge.

Thanks are due first of all to Paul Hoffman, who spent countless hours talking to me. He drove me around the Namibian desert and the Boston marathon course, welcomed me into his field camp, his home and his lab and sent me a continual stream of information by e-mail. He sought no influence over what I wrote.

Thanks also to Dan Schrag, who first introduced me to the Snowball. His ideas have been central to the theory, and he shared them with me repeatedly and very willingly.

Many researchers helped me explore the various Snowball sites around the world. Thanks to Tony Prave, Mark Abolins and Frank Corsetti for introducing me to the rocks of Death Valley in California. Ian Fairchild worked hard to set up the Snowball Earth workshop in Edinburgh, after his field trip to the Garvellachs was sadly scotched by the outbreak of foot-and-mouth disease. Joe Kirschvink invited me on his field trip to South Africa, and again to his lab in Caltech. (Thanks for Joe stories also go to Dave Evans, Ben Weiss, Kristine Nielson, Tim Raub, Curtis Pehl and many other members of Joe's irrepressible student family.)

ACKNOWLEDGMENTS

Dennis Thamm delayed his holiday for a day to show me around Mount Gunson mine. Jim Gehling and Linda Sohl drove with me to Pichi Richi Pass in South Australia, and Jim then gave me an unforgettable tour of the Ediacaran fossils in the Flinders Ranges. George Williams sent me copious papers covering his work on the South Australian deposits, in spite of his distaste for the Snowball idea. Kath Grey and Malcolm Walter talked to me about stromatolites before I headed to Shark Bay in Western Australia to see the living rocks there for myself. Guy Narbonne showed me the fabulous forms of the Mistaken Point assemblages, along with Bob Dalrymple and Jim Gehling. On the same trip Misha Fedonkin told me tales of the Russian White Sea, most notably while we sat in a Newfoundland late-night bar, and Bruce Runnegar gave me a whole new perspective of molecular clocks and pizza-shaped fossils. Thanks also to Ben Waggoner for his description of the flies at Arkhangel'sk.

Nick Christie-Blick told me his stories over dinner in Nevada and English tea in Newfoundland. Linda Sohl described the sheep incident and the noise that frightened kangaroos make, while the two of us were squashed together in the back of a van in Newfoundland. I shared a truck for three days with Martin Kennedy and Tony Prave in Death Valley and grilled them both further in the more genteel environs of Edinburgh.

Roland Pease enlisted me to present a BBC radio programme on the Snowball and provided me with copies of the interviews that we recorded. Mark Chandler explained about climate models and Jim Walker about seasons. Oliver Morton pointed out the significance of "outrageous hypotheses". Brian Harland tolerated three separate visits from me to his Cambridge home and office, the first when he had only just emerged from the hospital, having suffered a shattered knee. Ian Fairchild and Mike Hambrey told me more about Brian's early work, while fascinating insights into Paul Hoffman's background came from Terry Seward, Peter von Bitter, Erica Westbrook, Sam Bowring, Jay Kauffman and Dawn Sumner.

As for my own background, if it hadn't been for John Maddox and Laura Garwin, who took me on as a rookie at *Nature*, I would never have known about the wonderful world of earth science. Many thanks are also due to my colleagues at *New Scientist*, especially Alun Anderson, who hired me to do the best job in the world ("go out and look for stories that you find fascinating") and Jeremy Webb, who taught me more about writing and editing than either of us realize. The book began as a feature article for *New Scientist*, one of many that came from my insatiable thirst for stories about ice. The U.S. National Science Foundation sponsored me on two wonder-filled trips to Antarctica, and the Canadian Department of Fisheries and Oceans conveyed me to the Arctic Ocean to catch my first, gripping glimpse of what happens when the seas themselves freeze.

I wrote mainly in London at the British Library, where the staff is marvellous—particularly in Science 2 South. The town of Kirkcudbright in Scotland provided another haven for a while. I didn't manage much writing there, but did plenty of thinking and made twenty-four pounds of procrastination jam. In California I wrote in the beautiful Sausalito Public Library. And Niles Eldridge at the American Museum of Natural History in New York kindly granted me a desk for a month of writing there. I often stayed beyond closing time and felt a privileged shiver each night that I walked past the dinosaurs in the dark.

Many people read the manuscript in whole or in part and made helpful comments and criticisms. Thanks for this go to Robert Coontz, Richard Stone, Sarah Simpson, Michael Bender, John Vandecar, Helen Southworth, Rosa Malloy, Diane Jones, Jaron Lanier, Dominick McIntyre, Jeff Peterson, Edmund Southworth, Paul Hoffman, Doug Erwin and Jim Gehling. Any errors are, of course, my own. Particular thanks go to David Bodanis, who was there from the beginning with his encouragement and insights. He would save me a coveted booth at the British Library café and spend lovely long lunches there helping me find a way through my story. The edits he suggested were often painful, but invariably right. From David I also learned the two most important secrets for book writing: write every day, and always have a map.

My agent, Michael Carlisle, has given me unstinting support and encouragement. Alexandra Pringle, my editor at Bloomsbury, was full of enthusiasm from the beginning. I owe more thanks than I can say to Emily Loose, my editor at Crown, who encouraged me to write the stories I truly wanted to tell, with her intelligent suggestions on structure and her knack of teasing details out of me that I'd been tempted to gloss over. At one point, I read of a famous writer who challenged his editor about the importance of details over structure. God is in the details, the writer said. No, God is in the structure, his editor replied. When I asked Emily about this, she answered without hesitation. "God," she said firmly, "is in both."

I suppose I must have talked a lot about the Snowball over the past couple of years. Thanks for tolerance and encouragement go especially to Helen Southworth, Diane Martindale, Barbara Marte, John Vandecar, Jaron Lanier, Dominick McIntyre, Jonathan Renouf, Karl Ziemelis and Christine Russell. Thanks to Rachel Rycroft and

ACKNOWLEDGMENTS

Barbara Nickson for setting me on what—to my surprise, though probably not to theirs—turned out to be the right path after all.

Above all, thanks to my family: Rosa, Helen, Damian, Ed and Christian. They are a rock that never cracks or crumbles, no matter how bad the storms.

Out of whose womb came the ice?
And the hoary frost of Heaven, who hath gendered it?
The waters are hid as with a stone
And the face of the deep is frozen.

—*Book of Job*

SNOWBALL EARTH

PROLOGUE

BOSTON, 20 APRIL 1964

The weather was foul. Flurries of snow turned to wet sleet and slapped at Paul Hoffman's face as he warmed up on Hayden Rowe, just around the corner from the marathon starting line. The wind, he noticed, was gusting from the northeast. Not good. At six foot one, Paul made a large target for a headwind, and this one would be in his face for most of the twenty-six miles and three hundred and eighty-five yards to come.

Paul was tall for a distance runner, too tall, but at least he had the right sinewy build. He was lean and gangly and had just turned twenty-three. He'd always been athletic, but running suited him better than any sport he'd ever tried. Especially distance running. Paul liked his own company, and he didn't function particularly well in a team. He enjoyed the long solitary hours of training, and liked the feeling of pitting himself against the world.

Apart from his height, there wasn't much to distinguish him from the rest of the runners: his dark hair was neatly cropped, short back and sides, parted on the left; he had hollow cheeks and a long thin face. Though he was a star in his local athletics club, Paul was a nobody in these circles. He was anonymous, wedged in towards the back of the pack.

That was about to change.

On this chilly morning, Paul was feeling excited and nervous

I

in equal measure. He had never run a marathon in his life, not even in training, yet he had decided to start at the top. Boston is the oldest city marathon in the world, the race of legends. If you are a long-distance runner, this is the race that matters. Normally you'd build up to it, try a few lower-profile marathons, and work out your pacing. But Paul Hoffman had never been particularly interested in preambles. Up in his native Canada, he had pushed himself for mile upon lonely mile over the hills of the Niagara escarpment, and afterwards, as he neatly inscribed the latest times and distances in his running log, he calculated how his efforts would translate into a full marathon. Every time, he was looking for one figure: six minutes a mile. That was the mental cut-off. If he could sustain that pace for the whole length of a marathon, he'd have a most respectable time of just under two hours forty. As soon as Paul figured he could make the pace, he sent in his entry forms.

Now, on this cold, filthy day, the doubts were kicking in. What if he had guessed wrong? What if he ran out of resources shy of the finish line? How hard could he push himself at the beginning? *He'd never run the distance.* Even the crowd was unsettling. Paul had never seen so many spectators at a race. It was Patriot's Day holiday—the day commemorating the "shot heard round the world" that started the American Revolution—and despite the nasty weather, the citizens of Massachusetts were out in voluble force.

The race began at noon. From the outset, the course sloped downhill in a long looping curve, begging Paul to stretch out his legs and increase his speed. He resisted the impulse. "Keep breathing," he told himself. "Concentrate. How am I feeling? How am I *really* feeling?" His legs seemed fresh, and he was pacing well.

The headwind was blowing obliquely from the right. Some of the runners ducked behind each other to block it. But Paul kept open road ahead of him. If he ran behind someone, his stride felt cramped. He'd rather face the wind.

The field stretched out, and the leading pack pulled out of sight. Paul made his way steadily up to the second group, which contained six other runners. They swept into the town of Natick to a wave of applause from the crowd. "You're looking good," someone bawled at him. "Keep going!" A slender clock tower stood to the left, its hands pointing to 12:56. Paul had been running now for precisely fifty-six minutes. Natick was around the ten-mile mark, and at six minutes a mile the time should have been closer to one o'clock. "We're going too fast," Paul shouted to another runner alongside him. Neither of them slowed down.

On the other side of town, the group entered a road lined with bare winter trees and screaming fans. Thousands of students had poured out of the venerable Wellesley college buildings on the right. This was the Wellesley gauntlet, a tradition Paul could well have done without. A wall of screams and shrieks tore at his ears, as he shook his head irritably and ploughed on.

Soon the volume of the shouts dropped and he could recognize words again from the spectators. "Paul! You're fifteenth!" Fifteenth place. The ranking stunned him. This was better than he'd ever imagined.

There were four runners immediately ahead of him now. That meant there were ten people out of sight in the leading group. Ten people. Ten trophies. If Paul could outstrip the rest of his group and overtake just one of the leaders, he'd be in the top ten. From his first-ever Boston Marathon he would take home a trophy.

The road suddenly plunged downhill. At the bottom was Newton Lower Falls, and the first of the uphills. Paul knew he was good with hills. He took his chance, stretched out his stride, and pushed the pace, leaving the other four behind.

In . . . out . . . in . . . out . . . breathing was getting harder. No more holding back now. This was flat out, as fast as he could force himself to run. The yelling was increasing in pitch. "You're eleventh!" "There's no one behind you!"

Six miles to go, and Paul was entering the unknown. He knew his body would soon run out of sugar reserves and start burning fat. Sugar gives instant energy. But fat is a long-term energy store, not designed for hard, sustained effort. When you start burning fat, you hit the wall. Your legs feel suddenly like lead and your arms can no longer pump. This happens at different times for different people, but it's usually something over two hours. That's why marathons are so hard. Paul had no idea how long his sugar reserves would last.

Another hill appeared up ahead, the site of one of the Boston Marathon's most famous dramas. In 1936, reigning champion Johnny Kelley caught up here with a young Narragansett Indian, Ellison Myers "Tarzan" Brown, who had been leading the field. Kelley patted Brown's shoulder as he reached him. "Good race," the gesture said. "Goodbye." Brown responded with an instant surge. It was textbook running. When someone has just expended supreme effort to catch up with you in a marathon, they need to coast along for a while. Take off again, and you break their spirit. Also, in this case, their heart. Brown went on to win the race while Kelley limped in fifth. And the rise was immediately dubbed "Heartbreak Hill".

Paul was on his own now, with no runners in sight ahead or

4

behind as he set off up Heartbreak. At the brow, he passed two exhausted runners who had dropped back from the leading group. They were toast. He felt a surge of elation. That means ninth place! I'm in ninth place!

His legs were still working, there were no signs of cramp. But his body was screaming at him. All his concentration was taken up with ignoring the pain.

He turned left on to Beacon Street, heading into town on the long, straight stretch before the final few twists. Far up ahead blazed a Citgo sign marking the twenty-five-mile point. He was striving to reach it but the sign seemed no closer. Please let it be over soon. No runners in sight. Just the crowds yelling. "Paul! You're ninth! Paul! Keep going!"

When he finally reached the Citgo sign, a paltry rise in the road nearly finished him off. Somehow he forced himself to run on and make the right turn on to Exeter Street. He flicked his eyes back as he turned and saw another runner coming up hard behind him. Ahead, he caught his first glimpse of the leading group, and he summoned one last huge effort. He began to gain on the leaders, but the runner behind was gaining on him. In the end, all ran out of road. Paul crossed the finish line in ninth place, to a roar like he'd never heard before.

In his first marathon, Paul had run *two-twenty-eight-oh-seven*. He'd hoped for something like two hours and forty, a good ambitious target. But he'd wildly exceeded that. His marathon time was world class, less than fourteen minutes short of the world record. If he wanted to, he suddenly realized, he could make this his life.

Paul's summer was already planned. He was training to be a field geologist, and intended to go to the Arctic. The Geological Survey of Canada had offered him a place on a field party at

Keewatin, west of Hudson Bay. He'd been out in the field before, but this was the first time he would be a senior assistant, allowed to do independent mapping of the geology. To join the Survey full-time was Paul's dream, and this summer job could be the first step.

But everything suddenly seemed different. This was an Olympic year, so should he try for the Canadian Olympic team? Should he abandon his fieldwork for the summer and go to the games in September? He was already too late to have a serious chance of winning the Tokyo Marathon, but perhaps he should put his geology on hold and devote the next four years to training for Mexico City in '68.

He drove back to Canada that night, stiff-legged and aching, statistics spinning in his head. He'd run two-twenty-eight-oh-seven, an average of five minutes thirty-nine per mile. The world record was a five-minutes-ten pace. With intense training, could he shave twenty-nine seconds per mile from his time? Could he ever hope to match the world record? If he went to the Olympics, would he be in contention? Was he a world-beater? Could he *win*?

Back and forth he went in his mind, but he always came up against the same wall. Five minutes ten a mile. That was fast. He could run at that pace for two miles, five miles even. But not ten. And if he couldn't run ten miles at that pace now, even intense training wouldn't help him sustain it for a full marathon. He could run a good race in Tokyo, but he wouldn't win. Whichever way Paul looked at it, he couldn't see a chance at the Olympic gold.

That made the decision for him. If he couldn't win at the Olympics, he wouldn't go. He'd find another route to glory. Come the summer, he was camping in the Canadian high Arctic, beginning the painstaking process of piecing together the stories hidden deep in its rocks.

ONE

FIRST FUMBLINGS

This is an extraordinary time to be alive. Look around you, take in the intricate complexities of life on Earth, and then consider this: complex life is a very recent invention. Our home planet spent most of its long history coated in nothing but simple, primordial slime. For billions of years, the only earthlings were made of goo.

Then, suddenly, everything changed. At one abrupt moment roughly 600 million years ago, something shook the Earth out of its complacency. From this came the beginnings of eyes, teeth, legs, wings, feathers, hair and brains. Every insect, every ape and antelope, every fish, bird and worm. Whatever triggered this new beginning was ultimately responsible for the existence of you and everyone you've ever known.

So what was it?

Paul Hoffman, part-time marathon-runner, full-time geologist, and obsessive, intense seeker of glory, thinks he knows. He believes he has finally struck science gold. Now a full professor at Harvard University and a world-renowned scientist, he has uncovered evidence for the biggest climate catastrophe the Earth has ever endured. And from that disaster, according to Paul, came a remarkable new redemption.

SHARK BAY shows up from the air as a snag in the smooth coastline of Western Australia. Five hundred miles north of Perth, it lies just at the place where tropical and temperate zones rub shoulders. The area around the bay is a powerful reminder of how far we have come since primordial slime ruled the world. It is full of varied, vivid life.

This is one of the few places in the world where wild dolphins commune with humans, every day, regular as clockwork. At 7:00 A.M. each day a park ranger dressed in khaki uniform emerges from a wooden hut to focus a pair of binoculars on the horizon. Perhaps half an hour later, he'll spot the first dolphin fin. Somehow the word immediately spreads. Where there were only four or five people on the sandy beach, suddenly fifty or sixty appear.

Three harassed rangers do their best to marshal them into an orderly line. Everyone will get a chance to see the dolphins. No one will be permitted to touch them. No one must go more than knee-deep in the water. Another ranger deftly diverts the enormous wild white pelicans away from the beach by flicking on a water sprinkler. The birds flock around with gaping jaws—in this desert landscape, fresh water is irresistible.

The dolphins and their calves arrive. One of the rangers, a

wireless headset amplifying her voice, wades up and down in front of the spectators, introducing the dolphins ("This is Nicky and Nomad, Surprise and her calf Sparky") and reciting useful dolphin facts. The crowd surges into the water, like acolytes seeking a Jordanian baptism, their expressions beatific.

The dolphins are the crowd pullers—more than six hundred of them live here. But Shark Bay is also famous for the rest of its wildlife. The bay contains more than 2,600 tiger sharks, not to mention hammerheads and the occasional great white. The tigers show up in the water as streamlined shadows up to twelve feet long; often they are skulking beside patches of sea grass in the hope that dinner will emerge in the form of a blunt-nosed, lumbering grey dugong. Dugongs, or sea cows, are supposedly the creatures behind the mermaid myths, though I can't see it myself. They are too prosaic, placidly chewing away at the end of a "food trail", a line of clear water that they have cut, caterpillar-like, through the fuzzy green sea grass. They're exceptionally shy and rare, but here, among the largest and richest sea-grass meadows in the world, are a staggering ten thousand of them—tiger sharks notwithstanding.

Then there are sea snakes, green turtles and migrating humpback whales. And just a little to the north, where the tropics begin in earnest, lies Coral Bay—one of the world's top ten dive sites. *Come and dive the Navy pier! See more than 150 species of fish!* Also sea sponges and corals, brilliant purple flatworms, snails and lobsters and shrimp. And the vast, harmless whale sharks, the world's biggest fish. And on land there are wallabies and bettongs and bandicoots, emus and kangaroos and tiny, timid native mice.

There's everything in this region, from the wonderful to the plain weird. Evolution has been tweaking, adapting and inventing new forms of complex life for hundreds of millions of years, and here in Western Australia it surely shows.

But this is also a place where you can travel back in time, to see the other side of the evolutionary equation—the simplest, most primitive creatures of all. They come from the very first moments in the history of life, just after the dust from the Earth's creation had settled. And when these first fumblings of life appeared on Earth's surface, their form was exceedingly unprepossessing. Throughout oceans, ponds and pools, countless microscopic creatures huddled together in a primordial sludge. They coated the seafloor, and inched their way up shore with the tide; they clustered around steaming hot springs, and soaked up rays from the faint young sun. Dull green or brown, excreting a gloopy glue that bonded them together into mats, these creatures were little more than bags of soup. Each occupied a single cell. Each had barely mastered the rubrics of how to eat, grow and reproduce. They were like individual cottage industries in a world that had no interest in collaboration or specialization. They were as simple as life gets.

Although these primitive slime creatures have now been outcompeted in all but the most hostile environments, a few odd places still exist where you can experience the primeval Earth first-hand. The acidic hot springs of Yellowstone National Park, for instance, or Antarctica's frigid valleys. And here, in Western Australia, where countless microscopic, single-celled, supremely ancient creatures are making their meagre living in one small corner of Shark Bay: a shallow lagoon called Hamelin Pool. The pool's water doesn't mix much with the rest of the bay, and it's twice as salty as normal. Since few modern marine animals will tolerate so much salt, this is one of the last refuges of ancient slime.

THE SIGN pointing to Hamelin Pool is easy to miss, even on the desolate road running south from Monkey Mia. On the second pass I finally spot it, turn left, and bump along a sand track with scrubby bush to either side. For this first visit I avoid the restored telegraph station with its tiny museum and tea shop, and head straight for the beach. I want to experience primordial Earth without a guide.

There's an empty car park of white sand, with wattles and low-slung saltbushes clinging to the surrounding dunes, and a path threading through the bushes towards the sea. Though I've come to find the world's simplest creatures, the complexities of life are everywhere. From one of the bushes a chiming wedgebill incessantly reiterates its five-note melody. From another, a grey-crested pigeon regards me unblinkingly. The shells of the beach crunch underfoot; they are tiny, bone-white, and perfectly formed, and the bivalves that grew them are eons of evolution ahead of the simple creatures that I'm seeking. I step on to the boardwalk, which stretches like a pier out into the water. Each weathered plank of wood contains row upon row of cells that once collaborated in a large, complex organism. Signs on all sides show pictures of the slime creatures with smiley faces and cheery explanations of their origin. Flies buzz infuriatingly around my head, landing on my face to drink from the corners of my eyes. Black swifts swoop between the handrails, and butterflies the colour of honey, with white and black tips to their wings. Time travel is harder than it looks. The modern world is right here even in Hamelin Pool, and it's stubbornly refusing to leave.

I retreat to the telegraph station to plead with the ranger for permission to leave the boardwalk and wade out into the pool. He hesitates and then relents. "Go along the beach to the left," he says. "Don't step on the mats. Be careful." The mats he's talking about are one of the signs of primeval Earth. They are slimy conglomerates of ancient cyanobacteria, and they grow painfully slowly. At the beginning of the last century, horse-drawn wagons were backed into the sea over the mats, to unload boat cargo. A hundred years later the tracks they left are still visible as bare patches in the thin black sludge. An injudicious footprint here will last a long time. I promise to watch my step.

I return to the beach and this time walk carefully towards the water's edge. More striking than the ubiquitous patches of sludgy, foul-smelling bacterial mats are the "living rocks" in between. These strange denizens of Slimeworld are everywhere, an army of misshapen black cabbage heads marching into the sea.

The ones highest up the shore are now nothing more than dead grey domes of rock, shaped like clubs, perhaps a foot tall. They once bore microbial mats on their surfaces, but these have long since shrivelled, abandoned by the receding water. Closer to the Pool's edge the domes are coated with black stippling that will turn to dull olive green when the tide washes over them. Most of the stromatolites, though, lie in the water, stretching out as far as I can see. Between them the sand is draped with black-green mats of slime, and chequered with irregular patterns of sunlight as the waves ripple overhead. I wade up to my knees among these strange formations, basking in the sunshine. There is nobody else in sight.

The living rocks of Slimeworld are called "stromatolites", a word that comes from the Greek meaning "bed of rock". Though the interior of the stromatolites is plain, hard rock, their outer layers are spongy to the touch. Here on the surface is where the ancient microbes live. They're sun-worshippers: by day they draw themselves up to their full filamentous height—perhaps a thousandth of an inch—soak up the sun, and make their food; by night they lie back down again. The water that surrounds them is filled with fine sand and sediment stirred up by the waves. Gradually this sand rains down on the organisms, and each night's bed is a fresh layer of incipient rock. The stromatolites are inadvertent building sites; the sticky ooze that the organisms extrude acts as mortar and the sand acts as bricks. Every day, as the microbes worm their way outward, another thin layer of rock is laid down beneath them.[1]

It's a slow process. Stromatolites grow just a fraction of an inch each year. The ones in Hamelin Pool are hundreds of years old and would be astonishing feats of engineering, had they been created by design. For these micro-organisms to erect a stromatolite three feet high is like humans building something that reaches hundreds of miles into the sky, and scrapes the edges of space. I wade a hundred yards, two hundred yards offshore, and the slope is still so gentle that the deepening is barely perceptible. Mercifully, the flies and butterflies have dropped back, and the birdsong is out of earshot. At last I begin to feel that I've travelled back to life's earliest days.

HAMELIN POOL'S mats and stromatolites look utterly alien, but they were once ubiquitous. Time was, this scene of stromatolites and stippled microbial mats would have greeted you everywhere you went. Forget dolphins and wallabies. This is how the Earth looked for nearly three and a half billion years. The imprints of the stromatolites and their mats show up still wherever sufficiently ancient rocks poke through to the Earth's surface. I've seen them in Namibia, in South Africa, in Australia and California. They are sometimes dome-shaped like these in Shark Bay, sometimes cones, sometimes branching like corals. There are places where you can walk among ancient petrified stromatolite reefs, rest your feet on their stone cabbage heads, and see where they have been sliced through to reveal rings of petrified growth. And you can run your fingers over fossilized mats, which give rock surfaces the unexpected texture of elephant skin. This slime used to be everywhere, and now it's almost nowhere.

How did we get from there to here? This is at once a simpler and more powerful question than it seems. Of course, life took many separate evolutionary steps on its way from stromatolites to wallabies. It had to invent eyes and legs and fur and feet, and everything else that distinguishes marsupials from slime. But there was one particular step that was more important than all the others, one that made all the difference.

The step was this: learning to make an organism not from just one cell, but from many. Though the first microbes on Earth were woefully unsophisticated, they did gradually learn new tricks to exploit the planet's many niches. But they all still had one thing in common. Each individual creature was packaged in its own tiny sac, a single microscopic cell. Then at some particular point in Earth's history, everything changed. One cell split into two, then four. From that time onwards, organisms could be cooperative, and above all their cells could specialize. There could be eye cells and skin cells, cells to make up organs and tissues and limbs.

For life, this was the industrial revolution. Forget the old cottage industries. Now you could have factories with production lines. Parcelling out tasks and specializing is always more efficient than trying to do everything yourself. And there are some things, wallabies for instance, that can only be made with a massive collaborative effort.

In just the same way, when organisms developed the ability to become multicellular, they gained a world of possibilities. Your body is made up of trillions of cells. Every hair is packed with them. You shed skin cells whenever you move. Your blood cells carry energy around your body, to feed the organs made up of still more cells. This multiple identity is the one criterion that's vital for any complex creation. Every dolphin and dugong, every shark, pelican and wombat depends for its existence on that crucial leap from one cell to many. This was the point when simple slime yielded its pre-eminence to the complex creations that heaved their way out of the sludge and started their march towards modernity.

But why did it take so long? The Slimeworld lasted for almost the whole of Earth history. Let's put in some numbers. Our planet had been around for 4 billion years before the first complex earthlings emerged from the ooze. That's nearly 90 per cent of Earth's lifetime.

Four billion years is an insane amount of time, almost impossible to contemplate. There have been many attempts to capture this spread of time in ways that we can comprehend. If the history of life on Earth were crammed into a year, slime would have ruled through spring, summer and autumn, continuing well past Halloween into the beginnings of winter. If it were squeezed into the six days of creation, slime ruled until six o'clock on Saturday morning. If it stretched over a marathon course, slime would have led the field past the twenty-three-mile mark.[2]

But my favourite image is this one, borrowed from John McPhee.[3] Stretch your arms out wide to encompass all the time on Earth. Let's say that time runs from left to right, so Earth was born at the tip of the middle finger on your left hand. Slime arose just before your left elbow and ruled for the remaining length of your left arm, across to the right, past your right shoulder, your right elbow, on down your forearm, and eventually ceded somewhere around your right wrist. For sheer Earth-gripping longevity, nothing else comes close. The dinosaurs reigned for barely a finger's length. And a judicious swipe of a nail file on the middle finger of your right hand would wipe out the whole of human history.

Stephen Jay Gould set the discovery of these vast stretches of Earth time in a long line of findings that put humans firmly in our place.[4] Galileo, said Gould, taught us that the Earth isn't the centre of the universe. Darwin, that we're just another kind of animal. Freud, that we're not even aware of most of the things going on in our own heads. And geologists have now discovered that the Earth reached late middle age before we were so much as a glimmer in its eye.

Though we humans are certainly complex, also clever, perhaps even the highest form of life that Earth has so far produced, we're nothing like the most natural earthlings. Measured in units of staying power, Earth's first, most primitive experiment in life was also its best. With simple individual cells, nothing complex, nothing flashy, each creature out for itself, life had found a supremely winning formula. Why should it ever change?

That's the question that has plagued complex, clever, thinking, adaptable humans since they first uncovered this bizarre history of life. Earth looked set to stay locked in slime for ever. Why did complex life appear at all, and why did it wait to emerge until that one point in time, just a few hundred million years ago, nearly at the end of the marathon, somewhere near your right wrist, late in the Earth's middle age?

To answer this, Paul Hoffman has seized on an idea that was first proposed sixty years ago, and was then dropped, half-heartedly resurrected, and dropped again several times over the intervening years. There's nothing half-hearted, however, about the resurrection Paul has now effected. He's marshalled new evidence, restored and amalgamated old ideas, and employed fierce argument to persuade the people around him. According to Paul, life's richness, diversity and sheer overwhelming complexity arose from a mighty catastrophe. It's called the "Snowball Earth".

FIRST CAME the ice. It crept from its strongholds at the North and South Poles, freezing the surface of the ocean, spreading gradually over the Earth's surface. A blue planet inexorably made white.

Individual crystals of ice first appeared in the sea like tiny floating snowflakes. They were smashed together by wind and waves, their fragile arms broken and their debris turning the seawater into a greasy slick. The surface thickened and froze into a thin transparent layer. As this layer thickened, it grew grey and then opaque from salt and air bubbles that filled its inner voids. In some places the greasy ice congealed into large round pancakes, with raised edges like giant lily pads, where they bumped and smashed against each other. And, a nice touch this, the fresh young sea ice grew a coating of frost flowers, each one the size and shape of an edelweiss.

Sea ice bends. Unlike freshwater ice, which can shatter like glass, the ice that forms first on the surface of the sea is elastic. When you try to walk on it, your legs unexpectedly buckle. But as it thickens it becomes reassuringly firm, like solid rock.

Though sea ice is grey when it first forms, it whitens year by year as its brine drains back into the sea. Even grey young ice is often dusted with white snow. But a frozen ocean is far from monochrome. Gashes of open seawater, created as the pack ice is ripped apart by wind and weather, expose the deep turquoise roots of the floating sea ice. And the dark ocean reflects in the clouds, streaking them the colour of a bruise. "Water sky" this is called, and polar sailors have long used it as a clue for where to point their ship next as they navigate perilously through the pack.

Where waterways have frozen over, the ice is smooth and level. Where the edges of an old water wound have been cauterized together again, untidy piles of ice blocks are an astonishing bright blue. Ice cracks suddenly like a whip. Sometimes pack ice groans and creaks as the wind crams floes together or prepares to break them open. But for the most part, the frozen polar oceans are shrouded in silence—eerie and absolute. There is no scene more alien on Earth.[5]

For perhaps a few thousand years, the white menace stole unheeded towards the equator. Earth's primitive life-forms had neither the eyes to see the encroaching ice nor the wit to fear it. Most of them lived their dull lives in a band around the Earth's waist, and as the ice advanced steadily from the far north and south, they bathed unconcerned in the warmth of their shallow, equatorial seas.

An occasional storm might have whipped up waves near the

shore. Perhaps the surf tore at the rubbery microbial mats that coated the seafloor and sprayed nearby rocks with scraps like soggy chicken skin. Stromatolites built up their stone reefs, layer by microscopic layer. Geysers blew. Rain fell. The sun shone again. There was no hint of the devastation to come.

But when the ice reached the tropics, its slow creep became a sprint. In a matter of decades, it engulfed the tropical oceans and headed for the equator.

Ice spread out from shallow bays and grew first a skin, then a carapace over the oceans. It clung to the beaches and scraped the mats on the seafloor. In some places this shell was still thin enough to crack and seal again. In others it was thousands of feet deep.

For a few hundred, perhaps even a few thousand years after the oceans were capped with ice, the land remained bare. But ice began to accumulate, gradually, in the thin air of mountain ranges, creating great frozen rivers that flowed down to fill the surrounding valleys. In the end, the whiteout was complete. Earth's surface looked like the frigid wasteland of Mars, or one of Jupiter's ice-covered moons. Sunlight bounced off the bright surface and was dazzled back into space. The mercury hit a staggering minus 40 degrees C. (Or it would have, except that at those temperatures mercury itself would have frozen.) There was little wind or weather of any kind. Clouds, by and large, disappeared, save perhaps for tiny ice crystals high in the atmosphere, which scattered sunsets into strange, lurid colours, blue and green, rimmed with vibrant pink. No rain fell and little snow. Every day brought silent, unremitting cold.

The Snowball wasn't just another humdrum old "ice age" like

those from more recent eras. The events we call ice ages were merely brief cold blips in an otherwise fairly comfortable world. There was ice in New York then, but none in Mexico. If you were in northern Europe during one of those ice ages, you shivered. But if you were in the tropics, you scarcely noticed.

Instead, Paul's Snowball was the coldest, most dramatic, most severe shock the Earth has ever experienced. It was the worst catastrophe in history. For perhaps a hundred thousand centuries, Earth was a frozen white ball, desolate and all but lifeless.

To the microbes, the Snowball must have seemed like the end of the world. Some survived, of course—they must have, or we wouldn't still see them today. Perhaps they huddled for warmth around undersea volcanoes. They might have survived near hot springs, or found fissures and cracks in the sea ice where the sun's rays could slip through. But for many, perhaps most, the Snowball was disastrous.

Eventually the Snowball empire began to founder. Volcanic gases gradually built up in the atmosphere, trapping the sun's heat and turning the air into a furnace. After millions of years of stasis, the ice finally succumbed, melting in a rapid burst of perhaps just a few centuries. Temperatures now soared to 40 degrees C. Intense hurricanes flooded the surface with acid rain. Oceans frothed and bubbled, and rocks dissolved like baking powder. Earth had leapt out of the freezer and into the fire.

There was at least one more of these Snowball-inferno lurches, and there may have been as many as four. But at the end of them all, after the last of the Snowballs and its attendant hothouse finally faded, some 600 million years ago, came the most important moment in the history of evolution. The rocks that appeared immediately afterwards bear fossils showing the first stirrings of complex life. Out of the ice and the fire that followed had come the complexity that we see around us today.

THIS IS Paul Hoffman's vision, and he is enchanted by it. Most other geologists are horrified. Accept his story, they say, and you have to reconsider everything you thought you knew about the workings of the world. Geologists are taught from an early age that the Earth is a slow and steady place. The past looked pretty much like the present. Change happens only very slowly, nothing is terribly extreme. True, there have been a few occasions where they have been forced to admit, somewhat grudgingly, that this picture falls short. The idea that an asteroid came from space to wipe out the dinosaurs was once derided, but is now widely accepted. OK, the argument goes, so the occasional extraterrestrial calamity can rock the Earthly boat. But broadly speaking, the geological picture of Earth's history is a settled, safe, comfortable one.

Compare that to Paul's picture of the Snowball. A global freeze. A planet that looked more like Mars than home. Ice *everywhere*. And then a sudden lurch from the coldest to the hottest that the Earth has ever been. Every way you look at it, his Snowball stretches the bounds of decency. It's as extreme and catastrophic as they come.

Small wonder, then, that the Snowball has become the most hotly contested theory in earth science today. Paul Hoffman, though, is resolute. He is the chief champion of the theory. By argument, evidence, and brute force of personality, he is determined to win over the unbelievers.

Paul is an obsessive man espousing an extreme theory. If he is proved right, we'll have learned something important about where we all ultimately come from. But there's a darker side to Paul's theory. He has uncovered behaviour in our planet that's unsettling in the extreme. If his vision is true, Earth can experience sudden lurches in climate that are more violent, and deadly, than anyone had ever imagined, and such catastrophes may well happen again.

TWO

THE SHELTERING DESERT

In the autumn of 1994, Paul Hoffman was back in Boston, nearly thirty years after he'd won his marathon trophy there. Though he'd continued to run marathons in his spare time, Paul had spent most of the intervening years sticking to his geological guns. He had acquired that most essential of accoutrements for the male geologist, a beard. His hair was more unruly these days. Thick and white like a goat's, it sprang up in surprise from a high forehead that was lined from too many days spent outdoors. Now fifty-three years old, he wore a pair of round wire glasses and was widely considered one of the top geologists of his generation. He had been elected a member of the prestigious National Academy of Sciences, had won countless awards, and written classic academic papers. He was back in Boston not as a callow nobody running the marathon, but as a full professor in the Department of Earth Sciences at Harvard University.

Paul had made it, then, into the ranks of world-class science. But still he wasn't satisfied. A cloud hung over him that he was desperate to shake off. After thirty years of fieldwork in the high Canadian Arctic, Paul had been abruptly forced out. He had picked a fight with the head of his home institution, the Geological Survey of Canada, and paid a high scientific price. A high emotional price, too. He had felt more at home working in the far north than anywhere else in his life, and now he was banned from returning there. When the blow first struck, he felt humiliated and lost. Now, two years later, he was arriving at Harvard with as much to prove as ever.

About this time, Paul's alma mater, McMaster University, contacted him as part of a survey of distinguished alumni. They asked what he'd like to be remembered for, and Paul replied without hesitation. "Something I haven't done yet," he said.

PAUL HAS been fascinated by minerals since he was nine years old. Next door to his elementary school in Toronto was the Royal Ontario Museum, and as a child he used to haunt the place. On Saturday mornings the museum held field naturalist classes, and Paul signed up with enthusiasm. The first year, the class studied butterflies. The second, fossils. But in the third year, Paul found himself studying minerals. They were perfect. It suited the atavistic urges of a young boy to acquire sparkling, shiny crystals of hornblende, quartz and fluorite, to hoard them and examine them, to try to obtain one of *everything*.

There were plenty of samples to be discovered in the rocks around Toronto, and always the chance of a new crystal, a rare crystal, a bigger, better sample than one Paul already had. And then the bargaining would begin. What did you find? What do you have that I haven't got? What have I got that you're dying for? Perhaps I'll trade you.

Paul's life became a treasure hunt. At first his mother would drive him and fellow members of the mineralogy club to their sites, but as they grew older they went unaccompanied. They scoured the public records in Toronto for locations of old, abandoned mines, then set off on camping trips to find them. Or they travelled to existing mines, where they charmed the workers and won permission to poke around the waste dumps.

Once they visited a quartz mine where a single large cavern was crammed with spectacular crystals, some milky, some as clear as water, some as long as your arm, and all gleaming like ice. From a mine in Cobalt, northern Ontario ("the town that silver built"), Paul brought back thin plates of native silver embedded in bright pink cobalt salts. During one morning of careful searching you could find ten or twenty ounces of silver among the rocks that had been tossed into the waste there. In the dump at a uranium mine, he found black cubes of uraninite set in a mass of pink and white calcite; also chunks of purple fluorite housing spectacular yellow needles of uranophane—calcium uranium silicate. Both of these uranium minerals are radioactive. Paul and his friends bought cheap Geiger counters from a scientific supply store in Toronto. They held the Geiger counters up to their finds and were thrilled by the staccato crackles that emerged. They weren't afraid of the radioactivity. As long as you're careful, as long as you don't spread the dust on your toast in the morning, you'll be fine.

Every fine weekend, Paul would head off to another mineral

site. He loved being outdoors. Continuing to collect minerals avidly throughout his teenage years, he evinced no interest in dating girls or following fashion. Instead he traded samples with the museum mineralogists, and swapped stories with the scientists at the University of Toronto. In minerals, Paul thought he'd found his métier.

But in 1961, during his freshman year at McMaster University in Hamilton, they turned out to be a major disappointment. The study of minerals mostly happened in a lab, it seemed, where you spent your day leaning over a desk, measuring the distances between spots on photographic film. Paul wanted to be outdoors, back on a treasure hunt. He wandered from the Mineralogy Department along to Geology, which sounded like the next best thing. Were there any opportunities for the summer? They sent him to the Ontario Department of Mines in Toronto, where the austere director, J. E. Thompson, looked him over and decided to take him on board. "Take the overnight train to Sioux Lookout on May tenth," Thompson told him. "Bring a good pair of boots."

Sioux Lookout was a tiny town surrounded by the ribbon lakes and dense forests of northern Ontario. Paul took both the train and the boots and soon found himself on a bush flight out into the wilderness. The lake shores were gorgeous, but the interior was a treacherous, forbidding place of dense bush and swampy ground. To reach the outcrops of rock hidden among the trees, you had to take a compass bearing and then fight your way through the undergrowth, counting paces to see how far you had travelled. The four members of the field party lived out of two canoes. Each morning they struck camp, stowed their tents and gear in the canoes, and then paddled on to a new site.

THE SHELTERING DESERT

It was a bad year for forest fires, and sometimes the smoke grew so thick that the researchers could scarcely breathe. And then, several weeks into the trip, David Rogers, the party leader, felt a crippling pain in his gut, which he quickly realized must be appendicitis. There was no point in waiting for the weekly supply plane. Paul stayed with Rogers while the other two paddled north through the night for help. Eventually their route intersected with a railway line, and they managed to flag down a train. An intercontinental train takes a long time to stop, even after the driver has seen two young men frantically waving from the bush, and has slammed on his brakes. The driver and his precious radio finally came to a halt several miles down the track. A hasty call summoned a bush plane to pick up the patient and whisk him away to a hospital. Geologists are tough. Three days later, sans appendix, David Rogers was back in his canoe, in the field.

That summer, Paul spent more than four months canoeing and traversing and mapping the rocks. This, he felt, was the life. He'd camped before, plenty of times. He'd even been in northern Ontario with his parents, on holiday. But this was camping for *work*. Every day there was a new site to explore, every day a new set of rapids to run. Paul Hoffman was hooked. Geology seemed to be exactly what he was looking for, and when the next summer came around, Paul was eager for more. But this time he wanted to go somewhere different. Sioux Lookout was great in its way, but it wasn't the true North. Paul was hankering after remoteness. What he really wanted was the Arctic.

Scratch a geologist and, under their skin, almost invariably, you'll find a romantic. They will often be gruff about the landscape they work in. They are usually matter of fact about the rocks and how they interconnect. But try asking why they've

27

chosen to spend their lives working in this particular place or on that particular terrain, and that's when the stories start to slip out.

When Paul was eight years old, just before he started with his mineral obsession, he heard a CBC radio drama about the last trip of the Arctic explorer John Hornby, an eccentric Englishman who had lived precariously in the Canadian Northwest Territories during the early 1900s. Hornby was quixotic, even by the extraordinary standards of the place and time. His eyes were an intense, piercing blue, and he refused—for luck—to travel with any man whose eyes were brown. Though his hair and beard were wild, he spoke with a soft, expensively educated accent. He was barely five feet tall, but his toughness was legendary. Once, so the stories go, he trotted for fifty miles beside a horse. Another time he ran a hundred miles in twenty-four hours, for a bet.

Hornby used to boast that all he needed, for a trip of any length, was a rifle, a fishnet and a bag of flour. He would take absurd risks, venturing into the barren lands again and again with scarcely any provisions. Finally, in 1926, he pushed the odds too far. He decided to spend winter in the remote Thelon River valley, a few hundred miles south of the Arctic Circle. With him he took his eighteen-year-old nephew, Edgar Christian, and an Edmonton man, Harold Adlard. Sometimes rowing, sometimes portaging, the party would take their hefty, square-sterned canoe across Great Slave Lake and eastwards to the Thelon River, where they would build themselves a log hut for the winter.[1]

The timing was crucial. This far north, winter would be excessively harsh. By November, thick ice would coat the lakes and rivers, and deep snow would smother the hut and its environs. After that, there would be little wildlife at large, and few opportunities for hunting. Yet Hornby, true to form, was taking few provisions. His entire plan relied on gathering meat from migrating caribou as they passed the Thelon River on their way south for the winter. If he missed the caribou, all would be lost.

Hornby, however, seemed to feel no urgency. He left several notes en route, stuffed into tins and marked by stone cairns. "Travelling slowly," one reported laconically. "Flies bad." And in another, left around 5 August: "Owing to bad weather and laziness, travelling slowly. One big migration of caribou passed."[2] By the time Hornby's party reached their wintering site sometime in October, most of the caribou had gone.

The party's attempts to stave off hunger grew increasingly desperate. They managed to trap a fox here, a hare there, sometimes a few scrawny Arctic ptarmigans. By early December, Hornby was reduced to digging up frozen blood from the site of an old caribou kill. It made, Christian wrote in his diary, "an excellent snack". Every day the party set traps. Every day now the traps were empty. "Got nothing but damned cold," Christian wrote on 18 February. And on 23 February, "this game of going without grub is Hell". Soon they were pounding old bones to squeeze out any sustenance, and scraping hides for fragments of meat.

By now all three were far too weak from hunger to attempt escape. They were hundreds of icy miles from the nearest humans, and in their poor condition, that distance might as well have been thousands. Edgar Christian remained touchingly optimistic in his diary. "We can keep on till caribou come North and then what feasting we can have," he wrote on 26 March. But Hornby died on 17 April, and Adlard on 3 May. Christian himself finally succumbed to hunger at the beginning of June, just days before the caribou were due to return. Two years later the Royal Canadian

Mounted Police discovered Christian's diary and the three bodies. Christian had laid Hornby and Adlard side by side and covered them as best he could. His own body had fallen from its bunk and broken on the floor. The silver watch in the breast pocket of his shirt had stopped at 6:45.

When Paul heard a dramatized radio production of Christian's diary, this extraordinary story struck a chord. He had just seen John Mills's portrayal of the doomed explorer Robert Scott in the stirring adventure movie *Scott of the Antarctic*. Scott embarked with a small band of followers for a daring adventure at the opposite end of the world. He had hoped to conquer the South Pole for England, but his expedition, too, was disastrous. When he and his men arrived at the Pole in January 1909, they were horrified to see a Norwegian flag already flying there, courtesy of their arch-rival, Roald Amundsen. The air at the Pole danced with tiny crystals of ice, "diamond dust", which cast bright rings of light in halos around the sun. But Scott's mood was black. "Great God!" he wrote in his diary. "This is an awful place and terrible enough for us to have laboured to it without the reward of priority."[3]

There was worse to come. On the return journey, Scott and his men gradually succumbed to the appalling weather. First one perished; then another famously walked out to his death in a blizzard. Finally, Scott and his remaining two companions starved to death, trapped in their tents by another blizzard, just eleven miles from a food depot.[4]

Like young Edgar Christian, the polar adventurers left diaries and letters from which *Scott of the Antarctic* quoted liberally. Scott's was particularly rousing. "Had we lived," John Mills's Scott intoned stentoriously at the end of the movie, "I should have had a tale to tell of the hardihood, endurance and courage of my companions which would have stirred the heart of every Englishman."

Scott and Hornby embodied the tragic heroes of fairy tales. Something about their stories tugged at the young Paul Hoffman. The two became confused in his head. He pictured Scott vainly seeking out caribou while icebergs crashed around him in the Canadian lakes. His eight-year-old mind retained only the haziest of details from these tales, but the romance of the planet's icy extremes took firm hold. One day, he'd decided, he would go north for himself.

So, in his sophomore year at McMaster, he began to ask around. The Arctic, he was saying. How can I get to the Arctic? For that, it turned out, he needed to approach the Geological Survey of Canada, an august, government-funded institution that sends geologists prying and poking at rocks in the remotest, most inaccessible locations. Paul took himself off to Ottawa, to sign up. Two months later he had won a place at a Survey field camp on the borders of Great Slave Lake in the Northwest Territories, just a short canoe ride from where John Hornby had suffered that last bitter winter.

Paul nearly blew it. Only three days after he arrived, he was horsing around, practising shot-put and discus using the rocks from thereabouts. One false move later, he had sent a discus of Yellowknife slate slicing through the tent belonging to the field party's leader. Fortunately, its owner was not yet in residence. A hasty but meticulous stitching job and a surreptitious switching of the tents got Paul off the hook and allowed him to stay for the rest of the season. That was all it took. The Arctic drew Paul as nothing ever had before. He was to return almost every year for the next three decades, until his contact with the Arctic—and the Survey—was unexpectedly and bitterly severed.

In some ways the appeal of the North was immediate and obvious. Fieldwork in the high Arctic had the three things that mattered to Paul more than anything. His work was an intellectual pursuit, it involved strenuous physical labour, and it happened in a place that was as beautiful as Paul had ever experienced. But the wildlife, or more particularly the insect life, would have dampened the enthusiasm of most. True, there were three magnificent weeks in June when the ice was still breaking up on the lakes and the place was heaven on Earth. These were warm, sunny, peaceful days when anything seemed possible. But then the flies came.

The mosquitoes appear first. They are big and noisy and desperately annoying. They insert hypodermic needles into your skin, and the moment they bite, you can feel it. A few weeks after the mosquitoes come the black flies, smaller but more devious. They are master miners. They carve out a cavity in your skin, injecting you first with anaesthetic to prevent you noticing. The anaesthetic they use is a nerve poison. If you get a few hundred black fly bites quickly enough—within an hour, say—you begin to feel the effects of the toxin. You feel nauseous, can't concentrate, and lose your bearings. You struggle to hold a line of argument in your head.

Spend long enough in the Arctic, and you will develop your own definition of a bad fly day. According to Paul, a bad fly day is when you can hit your arm once and find a hundred corpses in your hand. On bad fly days, mosquitoes whirr and whine around your head in a dense claustrophobic cloud. Black flies crawl everywhere on your clothes and skin, and into every crevice. To avoid inhaling them, you have to breathe through your teeth. If you run your hand through your hair, it comes back greasy and bloody. At the end of a bad fly day, you empty your pockets of globs of dead and half-dead flies. They have crept up your wrist, down your neck, under your belt, down the top of your boots. On bad fly days you soak yourself with industrial-strength Repex, the repellent of choice. Repex doesn't keep the flies away, but it stops them from biting. It lasts two to three hours. On bad fly days you don't have to be reminded to reapply.

In the Canadian Arctic, between the fine few weeks of June and the return of winter in late August, every day that is not freezing cold or blasted with wind is a bad fly day.

And then there are the bears. The first time Paul encountered a grizzly, he had been out all day on a long traverse, walking twenty or thirty miles. He was heading back to camp around 11:00 P.M., walking north fast, straight into the setting sun, his baseball cap pulled down low to block the dazzling sunlight. Suddenly the bear appeared under the brim of the cap, coming for him at full speed. The animal was backlit, its body in shadow but surrounded by sunlight. The ends of its hair shone silver, and foam and saliva were spewing from its mouth in glistening arcs. All Paul could see was a bear-shaped halo of light and foam.

Man and bear stopped in their tracks and stared. Paul remembers thinking, *Stand still. Don't move. If it charges, fall on your right side and protect your right hand.* Paul's right hand was precious, his drafting hand, the one he used to draw his meticulous geological maps. But the bear didn't charge. Paul made the slightest movement to the right, and the bear turned and raced off to the left, to where her two cubs were waiting on a small knoll. She cuffed her cubs and hustled them away. A few seconds later the foaming, glistening vision had vanished.

After that, Paul kept a pair of running shoes beside his sleeping bag while he was in his tent at night. If something pawed at the side of the tent, Paul would throw a shoe to shock it, and then rush out to scare it away. There wasn't much danger if you were awake and could frighten the bear off. The real trouble was if a bear came to the camp while you were away. A black bear or a grizzly could tear a camp apart trying to find food, and that would be disastrous. If your tent was destroyed, you were at the mercy of the flies. All day long you were fighting flies. You had to have a refuge from them at night, or you'd go mad.

The only way out was to shoot any bears that persisted in returning to the camp. Paul had to shoot three bears over the years—two black and one grizzly. He hated every time. He was shocked how much red-blooded damage you could do with one little squeeze of a trigger. The grizzly was the worst. When it came into camp, it made an angry beeline for the helicopter. It had probably been buzzed by some idiot joy-riders and was out for revenge. That was hardly the bear's fault, but the helicopter was too precious to risk. As Paul loaded the rifle, he felt sick. Afterwards the same helicopter slung the grizzly's body back out into the bush.

Paul knew that he couldn't afford to let the flies or bears get to him psychologically, so he never did. Gradually he got used to them both. After a few weeks of building up tolerance, he found that new fly bites didn't swell so much. And if you could ever see beyond the buzzing, whirring, whining clouds that enveloped you, the landscape was vast, empty and gorgeous. There were no trees to block the skyline, just mile upon mile of rounded rocks and the boggy Arctic vegetation known as muskeg. Air and light both had a clarity that Paul had never experienced before. During the fleeting summer months of his field season, when the outer vestiges of winter melted briefly, there were ponds and pools and lakes of water everywhere. The ground squelched underfoot. The only sound came from the nesting birds, loudly defending their soggy territories and raising their young. Even they quietened down at night, although the midnight sun still shone then. All day long the sun was low on the horizon, and at midnight it reached its lowest point. Then the sunlight slanted most steeply of all, and the shadows were dramatic and long.

The short summer and continuous daylight put everything into overdrive. Eggs hatched into fledglings and then grew into birds that were ready to leave their nests in a matter of weeks. Flowers appeared in the scrapings of soil between rocks and among the spongy mosses and lichens of the muskeg and then vanished again almost immediately. Summer after summer Paul returned to the Arctic, now a fully-fledged geologist for the Survey. He strode out his rock contacts, mapped his terrains, noted down the rock types and their structures. He worked sixteen to eighteen hours a day. There was no sign that any other human had ever set foot there. Paul felt that he was master of the landscape.

The rocks he was working on were among the oldest in the world. They came from a catch-all time that geologists call the Precambrian, because it led up to the Cambrian period—which heralded one of the most significant changes in the history of the Earth. Naming time slices by what comes afterwards is a peculiar geological habit. More peculiar than ever in this case, because the Precambrian is much more than just a slice of time. The Precambrian lasted 4 billion years, covering nearly 90 per cent of Earth's entire history.

And yet, to geologists, this has long been considered the

Earth's Dark Age. Plenty may have been happening, but nothing was recorded for posterity. The rocks of the Precambrian are like the history books of Europe's medieval Dark Ages—a blank. What was missing? Fossils. Geologists rely heavily on fossils. One rock can look much like another, and to find out when exactly it was formed you need to look at the creatures that are locked inside. The Cambrian, roughly 550 million years ago, is the time when serious fossils first appear in the rocks. If you look at a section of rock from the beginnings of the Cambrian, you start to see real creatures with legs and teeth and armour plating, and you see changes in the fossils over time. In more recent rocks, dinosaurs appear and then vanish, making way for the fossils of mammals, fish and birds. Each has its own season and time, and each dates the rock that houses it. Fossils provide a ready-made timescale. They are like clocks left frozen in the rock. Every slice of geological time that comes after the Cambrian can be divided into periods and eons, according to the creatures that lived then.[5]

But before the Cambrian there were no fossils to speak of. And the few algae and the simple, single-celled creatures of Slimeworld that did bequeath their forms to the rocks stayed more or less the same for billions of years. Because of this, the rocks of the Precambrian just merge together into one long, undifferentiated mass. This was the geological Dark Age because there was simply no way to tell one time period from another.

Look at a standard geological timescale, a poster pinned on to every geologist's wall, and you'll see the Precambrian squeezed into a tiny, unimportant-looking box at the bottom. "This squashed period contains almost all of Earth history," the legend ought to say, "and yet we know almost nothing about it."

Paul was fascinated by the Precambrian. He felt sure that this long, mysterious period of time must contain important secrets about how the world works, and he dearly wanted to find them.

The first project Paul embarked on was trying to discover whether the continents behaved in the same way in the Precambrian as they do today. On geological timescales nothing stays still—not even continents. Over millions of years, continents skip and skate over the Earth's surface, some crashing together to throw up mountains, others ripping apart to create ocean basins. Paul wanted to know if this had always been true, even in the Precambrian. And if so, was the dance of the continents a minuet or a jitterbug? Were their movements carefully orchestrated, or a random bump and grind?

Gradually, Paul began to piece together the way the plates that would become North America moved during the Precambrian. Rather disappointingly, they looked just as random as in more recent times. They were clearly dancing a jitterbug, not a minuet. Still, he put his results together with geological maps from all over North America and began to trace exactly how the continent had formed. He discovered that most of the formation took place in a short, frenzied burst of activity around 2 billion years ago, when seven small plates crashed together and stuck in place. After eight painstaking years of researching this tale, Paul published a massive synthesis, which he called "United Plates of America".[6] The research required two skills: careful attention to detail and the sort of mind that can synthesize countless arcane facts into one overall, compelling picture. Nobody else in the world could have written it.

LIFE WAS good, even away from the rocks, during the long Canadian winters when Paul was forced back southwards to analyse his data and kick his heels. He was still running, and he had a new obsession to add to his life: music. As a teenager, Paul's attention had been caught by modern classical music, but in his junior year at college he was introduced to African-American music: modern jazz and pre-war blues; Ornette Coleman, John Coltrane, Eric Dolphy. He collected recordings voraciously throughout the seventies. Soon he had a thousand records, then two thousand and more. His opinions were characteristically forceful. Miles Davis and John Coltrane? Overrated. Charlie Parker and Dizzy Gillespie? Fabulous. They're the real musicians, the ones who deserve the credit they never fully get. And Louis Armstrong. His care with notes! His extraordinary musicianship! People were put off by Armstrong's stage persona. They thought he was an Uncle Tom. But Armstrong knew what he was doing. Every note, every rhythm was as precise as they come. Billie Holiday, a singer with true soul. Ella. Yes, she had a fantastic voice. Yes, great technique. But she was never compelling as a musical artist. She never connected emotionally.

Paul began to host a radio show, which was aired live on Wednesday evenings from nine to eleven. He played an eclectic mixture of jazz, blues, gospel, country-and-western, all from his own record collection. He talked about the history of the music, the particular idiosyncrasies of the musicians, the merits of different recordings. He talked about how to listen to the music, what to like, what not to like. His show developed a cult following, and Paul loved it.

Also, much to his surprise, Paul began to share his life with a woman he had known for years. He had first met her in the sixties at the home of his mentor at the Survey, a geologist called John McGlynn and his wife, Lillian. Erica Westbrook was a friend of the McGlynn family. She was often at the house when Paul visited. He hadn't particularly noticed her back then, nor she him: she was a scornful teenager when Paul was a driven young college student.

But things were different in 1976, when Erica offered to sublet his house in Ottawa for the summer while he was away in the field. Paul had just turned thirty-five. He had never even had a girlfriend. His lifestyle wouldn't allow it. He spent too much time out in the field, and when he wasn't in the far north, he lived for running, and music. He was, he had always felt, too self-focused to have time and attention for a family. Erica was tall, an inch or so taller than Paul himself. She had long, thick, black hair, a generous smile, and a habit of casting amused sideways glances. This time around, she found Paul intriguing. She laid a bet with a girlfriend about which of them would succeed in seducing Paul. Erica won.

Still, Paul didn't particularly see a future in the relationship. His attention remained focused entirely on geology in the North. Erica's response was drastic. She took a plane to Yellowknife and spent a long, fraught week in the Northwest Territories, in Paul's field site, in Paul's home turf. That was the only place she felt she could count on his attention. She spent the week arguing passionately. She wanted the relationship. She wanted Paul. Once again, she won.

It was never going to be easy. Erica was sociable and warm. She went on to work as a palliative care nurse. She was a people person. Paul was utterly focused on his work. Once, Erica's resolve nearly cracked. There had been a snowstorm in Ottawa and the garage roof had fallen on top of Paul's car. Paul's precious car. A shiny red Lotus Elan that he had bought to compensate himself when an injury left him temporarily unable to run. When Erica saw the roof and the car that morning, she realized something that drove her crazy. Paul had already left to go to the Survey. He must have walked past the garage. He must have noticed the roof. He had done nothing about it. The Lotus was *his car,* but yet again he had left everything to her. He hadn't even mentioned it. Erica raced back into the house and dialled her mother-in-law's phone number.

Dorothy Medhurst, Paul's mother, has always been a formidable woman. Paul describes her as a whirlwind. Everyone else describes her with very healthy respect, bordering on awe. She is tall and strong and passionate. She is an artist. She is uncompromising. At eighty-eight, she now lives alone in an isolated cabin thirty miles from Toronto. The cabin has no electricity, no telephone, and no running water. Dorothy prefers living that way. All of her children were raised to think for themselves, to embark on projects, to stay outdoors, not to be home until the streetlights were coming on. When Paul cried as a baby, Dorothy would put him in his crib out under the tree. "If you're going to cry," she told him, "go cry to the mosquitoes." Paul can still trace the pattern of those branches in his head. The home Paul grew up in was not a cuddly, touchy-feely one. There were no soft furnishings. The wooden floors were decorated with field lines for ball games. The walls were festooned with paintings. Paul called his parents by their first names. You judged people not by their blood connections but by their talents, and how they used them.

Even as an adult, Erica was rather afraid of Dorothy. But still, on that snowy day in Ottawa, she dialled the number and blurted out her frustrations. Dorothy listened thoughtfully. When Erica had finished, this is what she said: "I agree. It's not normal behavior. But you have to decide now if you're prepared to put up with it. Because it's not going to change." This was excellent advice. Erica knew immediately that Dorothy was right. Paul wasn't going to change. She had known from the beginning that he was focused and obsessive and intense. That was his strength as well as his weakness. It was the source of his charisma and also the thing that made her want to scream at him. If Erica wanted any part of Paul, she realized that she had to take all of him. She stayed.

Erica was a big influence on Paul. He sought her advice, and she helped to temper his ferocity. If she had been around on 6 July 1989, Paul would probably still be at the Survey, and would probably never have heard of the Snowball Earth. But she wasn't. She was away, and when Paul decided to let fly, there was no one to caution against it.

He had received an essay that enraged him. Ken Babcock, the new head of the Survey, had sent the essay to all employees. Babcock was a political appointee, and he had no truck with the academic-style freedoms of the Survey researchers. He criticized everything that had gone before. This isn't a university, he said, it's a service to our clients in government and industry. Researchers at the Survey felt that he talked like a bureaucrat, not a scientist. In his essay, he told them to "get back to basics". They should focus on the practical needs of government and industry rather than on esoteric academic research.

The essay, entitled "The Search for Excellence", infuriated many of the Survey scientists. They seethed at the implication that their work was deficient in some way because they were driven by academic curiosity. How dare Babcock suggest that their work was irrelevant just because there wasn't an immediate payoff? Many of them despised Babcock and his bureaucratic ways. They believed that turning the Survey into some kind of glorified consultancy would destroy its fine reputation. But they all held their peace, except Paul. Paul couldn't help himself. He wrote a memo to all his colleagues, taking issue with Babcock's entire stance. That might have been all right, had his penchant for sarcasm not prompted him to add a caustic rider at the end of the memo. "The search for excellence at GSC [the Survey]," Paul declared, "should begin at the top."

Paul's memo inevitably found its way into the offices of the local newspaper, the *Ottawa Citizen*. The *Citizen*'s report was immediate and gleeful. "Top Survey Scientist Rebukes Boss", the headline declared. "Controversy Rages at Elite Government Agency."[7] Paul very properly refused the paper an interview. Babcock, however, did give an interview, in which he pointed out rather sourly that in the private sector Paul's memo would have been grounds for dismissal. "He is truly one of our outstanding national earth scientists," Babcock told the *Citizen,* which then told the rest of Canada. "I suspect that his knowledge of the world of politics and management is less well developed."

Privately, Babcock was furious. He had been personally attacked by a subordinate and now the whole world knew it. His backhanded compliment in the newspaper was a sure sign that he wanted Paul out, but Babcock didn't sack Paul—he couldn't. Instead, over the next few years, Paul felt that he was becoming a nonperson. His funding requests were refused, and he was passed over for all privileges. He was even turned down when he requested unpaid leave to teach for a semester in the United States. Nobody was *ever* turned down for unpaid leave. Paul had spent many semesters away before without difficulty. He began to realize that he'd have to leave, but what he didn't realize at first was that this would also mean leaving the Arctic. Wherever Paul went in Canada, he quickly discovered, he would be unable to get funding to finish any work he had started under the Survey's umbrella. Paul's memo cost him more than he'd ever dreamed.

Today he makes light of it. "I left as I arrived," he declares, "fired with enthusiasm".[8] But at the time he was humiliated, bewildered and hurt. And what hurt more than anything else was being barred from his beloved Arctic. He desperately wanted to finish the work he had started there. He wanted to be back, mapping the terrain, hiking across the bleak landscape of the barren land, a place that felt more like home than anywhere else on Earth. For the second time in his life, Paul was walking away from something precious to him. He'd done it with the chance of Olympic glory, and now he was doing the same thing with the Arctic. This time, as before, he responded the only way he knew how. If he couldn't go back there and finish his work, he'd find something better. He'd find a new problem to solve, a new route to glory. He'd find something new to be remembered for.

But where should he go? Harvard University offered him a haven for his academic base, and he moved there gladly. But he needed a new field site, one with exposed rocks from the right time, the Precambrian. The rocks had to be fairly easy to reach logistically, and yet it was important they hadn't been excessively studied already; there was no point going somewhere that had already been picked over by other geologists. Paul needed somewhere fresh, a place where a great story was just waiting to be unearthed.

He toyed with one or two possibilities. Kashmir, perhaps, in northern India. Or maybe China would work. Then he found the perfect candidate. South West Africa had just become Namibia, having won its independence from South Africa two years earlier, with none of the presaged bloodshed. For decades before independence, scarcely any geology had been done there by outsiders, thanks to the military occupation by the South African Defence Forces. But newly independent Namibia was beginning to open up to the outside world. And most of the country was taken up with a vast, empty desert, full of exposed Precambrian rocks. They were younger than the rocks Paul had worked on before. Rather than 2 billion years old, they were more like 6 or 7 hundred million years old. That put them closer to the end of the Precambrian, closer to that strange point in time when fossils suddenly appeared out of nowhere. Perhaps they might even hold some clues about why life had suddenly lurched away from the simple world of primordial slime into the complexity that we see around us today.

Paul had other reasons to feel pulled towards Namibia. His father's brother, "Izzy", had lived and worked there. A few times during Paul's childhood, Izzy had travelled to Toronto full of tales, and the young Paul's eyes had shone. Namibia had been on Paul's list for decades. Africa, too. After geology, Paul's other obsessions were jazz and athletics. Africa had consistently supplied the masters in both departments. Namibia won on all sides.

PAUL HAD to start again from scratch in Namibia. He didn't even know where the best rocks would be, or which places he should concentrate on. He pored over aerial photographs of the terrain, and tried to pick out likely rock outcrops, looking for ones that he could drive to on bush tracks, or reach with a short enough hike from a possible camping ground; ones, too, where the rocks seemed to be slightly tilted, so he would be able to walk from layer to layer, up and down, back and forth in geological time, without having to scale a vertical cliff face. The balance was delicate, though. If this tilting had been accompanied by too much bucking and rippling of the Earth's crust, the rock layers would be too complex to interpret.

In June 1993, armed with a list of outcrops to visit, Paul set off for Namibia. The contrast with Canada was stark. There were, mercifully, no flies in the desert. But there were also no long, slanting shadows. Sunlight in Namibia glared fiercely overhead. The dark rocks would soak up morning sunlight, and for the rest of the day heat would pour relentlessly back out of them. At noon, when Paul wanted to find some shade after hiking and measuring for hours, there were no shadows to be seen. There was no midnight sun. Summer or winter, the days were frustratingly short, and an impenetrable darkness would fall abruptly each afternoon at 5:45.

There were also more people than Paul had ever worked among. Even in the desert, driving along a bush track, he would suddenly come upon a village of round mud huts clustered around a tall, rickety windmill that pumped water from the local well. Paul quickly learned to take a "landing fee" with him. The front seat of his truck was perpetually wedged with bags of sugar and tobacco to offer to the locals. He learned his first halting words of Afrikaans, how to say "please" and "thank you" and ask directions. Around those villages, it was easy to get lost. The ground had been grazed bare of dried grass and there was only baked mud, rutted with myriad tracks from animals and carts and, occasionally, the tyres of vehicles like Paul's dusty white Toyota. Everyone wanted to help. When Paul said thank you, the villagers would reply "Pleasure!" in a lilting, cheerful tone. If there was nobody to ask for directions, Paul would sometimes have to cast back and forth, trying this track and that until he finally found one that seemed to be heading the right way. Sometimes the tracks disappeared completely, and Paul had to divert down narrow gullies, setting his vehicle pitching and yawing over the rocks, three wheels on the ground, one in the air.

He had never worked using a vehicle before. In Canada, everything was by planes and helicopters, boats and boots. In Africa, he had to learn how to cut across a dried-out riverbed without getting trapped with his wheels spinning helplessly in the soft sand. The tricks, he discovered, were to let air out of the tyres until they were half-flat and could grip the loose surface more easily, and never, ever to touch the brake in mid-sand.

Gradually the memories of Canada began to recede, and Paul found himself relishing the harsh aridity of the African landscape: the sweeping valleys, the narrow, winding canyons and the disdainful kicks of the springboks that bounced out of the way of the Toyota. Though he would never deviate from his geology for anything approaching a tourist activity, he grew to enjoy seeing African wildlife in the wilderness, where it belonged. Sometimes as he drove he would see ostriches, their short tails bouncing as they jogged through the bush. He saw giraffes with their black velvet eyes and absurdly long lashes ("the most beautiful eyes in the world"); grumbling warthogs and baboons, herons and bustards and African grey parrots whose monotonous "waaah, waaah" sounded like a whining child. In the air there was a flash of yellow as a southern masked weaverbird emerged from its dangling sack of dried grass and mud. On the ground, termite nests towered, with their turrets and tubes and demonic spires, all vivid red from the rusty Namibian soil.

And there were rocks and rocks and more rocks, all unstudied and enticing. North of Windhoek rose the great pink granite intrusion of the Brandberg, Namibia's highest mountain, flanked with flat-topped, chocolate-coloured hills. These were the remnants of a plume of hot rock that had risen up from inside the Earth some 133 million years ago, when South America and Africa were last conjoined. The plume had flooded angry lava on to the plains of both continents, helping to rip them apart and open up the South Atlantic Ocean. Even the soil thereabouts was magma-dark, barely covered with a pale blond beard of grass. There were no bushes or trees, just squat *Welwitschia mirabilis* plants, with a woody root from which sprouted two flat, flailing leaves. Each plant is miraculously long-lived. Its leaves grow slowly and steadily, corkscrewing around each other for hundreds, perhaps even thousands, of years.

Farther north still, the Precambrian outcrops emerged from beneath the volcanic floods. When these rocks formed, more than 600 million years ago, Namibia was covered with a broad, shallow sea that left behind sandstones and mudstones, pink carbonates and dark grey shales. Peering closely at these rocks, Paul found the thumbprint whorls of the ancient stromatolites that had inhabited the Precambrian shores; he found sand dunes, beaches and lagoons, all now petrified and awaiting his notebook and hammer. And where the ancient seafloor once dipped towards a western ocean, barren, rocky hills stretched for mile after mile, their layers in places magnificently buckled and twisted into vast folds that dwarfed the tiny Toyota as it jolted along the canyon floors.

Paul was entranced. Such geological riches, yet scarcely any of them had been studied. Surely among all these outcrops he would find something important, some intriguing new insight into the history of the Earth.

He intended to continue studying the bump and grind of continental motions, just as he had in Canada. He was used to working alone or with just a few students in tow, but for his first field season in Namibia he brought along another Precambrian expert, Tony Prave, a researcher from New York. Tony is a wisecracking Italian American, a jobbing geologist in his late thirties who works hard and stays out of the limelight (and hence, broadly speaking, out of trouble). With his thick, dark, shoulder-length hair and bronzed face he could easily be mistaken for a Native American. His accent, though, is pure Hollywood mafioso. He has a wide, charming smile and slightly wary eyes.

Tony had spent most of his career working in Death Valley in California. He got to know the Precambrian rocks there by heart. But though he loved Death Valley, he jumped at the chance to go to Namibia with Paul. Paul was a famous field geologist. This was, Tony felt, the opportunity of a professional lifetime. Throughout that season, Paul, Tony and two graduate students moved from camp to camp and outcrop to outcrop in the remote Namib Desert. They mapped, climbed, hiked and studied, walking up gullies and down valleys, musing, interpreting and learning to understand Namibia's deepest history.

Paul affectionately called Tony "Pravey". The two of them got along brilliantly. They were both opinionated, both robust in their arguments, both fascinated by the rocks. Tony found Paul's methods exhilarating. Out on the outcrops, Paul's mind was like a steel trap. "Why? *Why?* What's it mean?" Paul would ask, rapid-fire, when Tony reported an observation. And then, as they headed to a new outcrop: "What would you predict? We're going over there now. What would you predict, Pravey?" Back at the camp, the two of them would stay up talking until eleven or twelve o'clock at night. Unlike the Arctic, where you could do geology at any hour, Namibia had long nights of enforced absence from the rocks. These were times to sit around the campfire and drink whisky and bandy opinions back and forth. It didn't seem to matter that they sometimes disagreed, that Paul adored baseball, for instance, while Tony hated it. They delighted in each other's company, and Tony basked in Paul's warm approval.

Eventually, inevitably, trouble started. Paul has never found friendship easy. He's charismatic, but also self-focused and intense. As a child, his relations with his fellow mineral collectors were civil rather than warm. In athletics, even when he was part of a team, he raced alone. And many of the geologists with whom he once worked closely are now scarcely on speaking terms with him. People like Tony. Tony and Paul no longer collaborate, or go on field trips together. They are no longer friends.

The problem arose towards the end of the field season, when Tony began to disagree with Paul's interpretation of the Namibian rocks. The issue was an arcane geological one, involving the details of exactly when Africa collided with South America. During the Precambrian, a narrow ocean separated these continental behemoths—they wouldn't actually hit until sometime in the Cambrian. But Tony became convinced that Africa was nonetheless beginning to sense the coming collision, and that its rocks had begun to buckle and bend in response. Paul, on the other hand, maintained that there was no sign in Namibia's Precambrian outcrops of the impending pile-up.

This disagreement started to sour their relationship, and by the time they returned to Namibia next season, the early warmth between them had ebbed away. Now, in the talk around the campfire, Paul was sarcastic about what he called "the Pravey hypothesis". "Oh, so what does the great Pravey say is going to happen tomorrow?" Tony remembers him asking. "What does the great hypothesis predict?"

"Paul's very competitive," Tony says now. "He's one of these people where if you take three steps, he has to take four. If you've mapped ten square kilometres, he has to map eleven. He's always got to be that little bit better, that little bit more intense." Tony had started to see the sharp side of this competitiveness, and he didn't like it.

The final straw came when Paul returned from a day's work mapping a narrow canyon. He arrived in triumph. "The Pravey hypothesis is dead," he declared when he re-entered camp. He had, he said, gathered evidence that conclusively disproved Tony's interpretation. The next day Tony hurried to the site, now dubbed by Paul "the Canyon of Contention". And there, among the rocks that Paul had mapped, Tony saw a whole jumble of fault lines. The rock layers had been mashed up beyond measure, long after they had formed. You couldn't use them to prove or disprove anything.

Tony saw red. He launched himself back to the camp, and confronted Paul head-on. They were eyeball to eyeball. Tony swore at Paul. Paul swore back. They began to yell, spittle flying from their mouths in their fury and making arcs in the air. The two students who were also on the trip looked on in horror. Then one of them decided to intervene. She was diminutive but tough, and she wedged herself between Paul and Tony. "You should be ashamed of yourselves!" she told them both. "Stop it!"

Her intervention worked. The shouting stopped, and Paul stalked off to his chair where he sat silently, staring into the darkness. The camp was subdued that night, and shortly afterwards Tony left Namibia.

Relationships among geologists are intense. By its nature, geology involves travelling with your colleagues to remote places, working long, hard hours in sparse conditions, living on top of one another and away from other people for weeks on end, having little contact with the outside world. Think of submarine crews, or Antarctic explorers. Think of throwing obsessive, opinionated people together in places that they can't easily leave. Their personalities become magnified. They bond or they break. Paul in particular has had plenty of fights like the one he had with Tony. Stand-up, screaming fights. He reacts furiously when confronted, and he holds nothing back. He will rage one moment, and ten minutes later act as if nothing had happened. But those on the receiving end are slower to forget. His reputation as a brilliant geologist has been tempered over the years with his reputation as a hugely difficult character. The people who still work with him are the few who know how to handle him. He can be rude, sarcastic and unpleasant. He's dismissive. He often makes people feel small. He knows this. He even makes a joke of it sometimes. "Everyone's entitled to my opinion," he'll say. And then, "Gosh, I'm awful. I don't know how I'd react to me."

And yet, when he compliments people, they feel good. They feel special. There is something about Paul that makes you want his approval. I have met former students of his who are now established geologists, tenured professors with great careers in top universities. They have all this, but they *still* care desperately what Paul thinks and says. If you ask them why, they shrug help-lessly. "I don't know," they say. Tony talks about the time before his fight with Paul as their "honeymoon", and the time after as their "divorce". Even now, years later, he gets a look of frustrated pain when he talks about their fight. Even now, he says this about Paul: "That man is so charismatic, if he'd been born two thousand years ago, he could have been Jesus."

PAUL SHRUGGED off his fight with Tony. He returned to Namibia the next season and the next, with a fresh batch of students to help map and measure and interpret. After that first flush of excitement, though, the outcrops were beginning to look dis-appointing. He had wanted to measure the timing of the conti-nental shifts, but the rocks in Namibia turned out to be useless for accurate dating. Still, Paul couldn't shake off the feeling that these outcrops held some important secret. He felt he was big-game hunting, but for what?

Then something began to nag at him.

Everywhere Paul went in Namibia, he spotted signs of ancient ice. He would be hiking up a gully and suddenly he would see a huge white boulder embedded in the grey siltstone. Siltstone is formed from an ancient seabed. Over time in the ocean, a fine rain of sediment lands gently on the seafloor and is gradually con-verted to rock. But a boulder had to be brought in separately from the shore. Something must have carried it out into the ocean and then flung it overboard. There were no ships in the Precambrian, and certainly no creatures capable of flinging anything. The culprit had to be icebergs. The boulder, a "dropstone", must have fallen from a melting berg up on the sea surface.

That wasn't all. As Paul looked more closely, he would see a medley of rocks appearing in the siltstone. Not a single boulder now, but countless pebbles and stones, all shapes, sizes and colours; fractured and rounded; pink, brown, tan, white and grey; granite basement, quartzite and carbonate. This mad jumble had somehow become bound up in the fine grey silt. Like the boulder, these rocks were interlopers. Something had gathered them up from mountains and gullies throughout Namibia. Something had bulldozed them down to the shore and on into the silty sea. The mix of multicoloured rocks stretched in every direction for hundreds of miles. Only one agent was capable of transporting so many different kinds of rock over such large distances: ice.

Paul recognized these ice-signs immediately. His mother's fireplace in her Canadian cabin was held up with two great chunks of pale stone packed with ice-borne pebbles. Paul had seen them every weekend and throughout the summer as a child.

He had also known for years that rocks like these show up all over the world. They can be found in the Americas, Asia, Europe—in fact on every single continent. And they all date from one particular point in time: the mysterious end of the Precambrian, just before the first real fossils appeared, just before life went complex and the Earth changed for ever. Paul had known all this even when he was working in Canada, but he had never really thought much about it. His mind had been fully occupied with his work on the shifting of continents.

Now, though, faced on all sides with the Namibian ice rocks, Paul started to wonder just why they were there. Why they were *everywhere*. You expect to see ice at the North and South Poles. But to find signs of ice on every continent seemed extraordinary. And the ice rocks in Namibia came with an extra mystery. They appeared in the middle of rocks that had clearly been formed in warm, tropical waters. What was ice doing in the tropics? And why did it appear there at that crucial moment in the history of life? Was it a coincidence? As he probed and pondered over the Namibian ice rocks, Paul grew more and more intrigued. He was haunted still by the sense that Namibia held some extraordinary story, just waiting to be told. Could these strange rocks be the key? Forget the motions of ancient continents. Now all he wanted to know about was ice.

What Paul didn't know yet was that the ice rocks brought nothing but trouble. For decades they had been grabbing the imagination of geologists without revealing their secrets. There was always some reason why ice in many of these places simply had to be impossible. Until now, everyone who had tried to explain the ice rocks had faltered. On the way, though, they uncovered clues that would prove vital for the Snowball story.

THREE

IN THE
BEGINNING

"Polar exploration is at once the cleanest and most isolated way of having a bad time that has ever been devised."[1] So wrote Apsley Cherry-Garrard, one of Scott's companions from the doomed South Polar expedition. It wasn't just the cold, or even the danger, that made early polar travel miserable, but the sheer physical effort of trudging over the snow for day after day, dragging *everything* on a sledge behind you. Henry "Birdie" Bowers, one of Scott's strongest and toughest men, called this the most backbreaking work he had ever come up against. "I have never," he said, "pulled so hard, or so nearly crushed my inside into my backbone by the everlasting jerking with all my might on the canvas band around my unfortunate tummy."[2]

These painful endeavours weren't confined to adventurers. If you were a geologist in the 1940s with a penchant for studying rocks in icy places, man-hauling was essential. There were no heli-copters, or snowmobiles. To reach remote, unstudied outcrops in the centre of any white wasteland, you had to load up all your equipment—tents, food, cooking gear, fuel—harness yourself to the sledge, and *heave*.

For Brian Harland, a geology professor from Cambridge University, the effort was worth it. Brian is famous for his precise probing into the Earth's past. He put together the definitive "Harland timescale", a chart of neatly coloured rectangles that divides geological time into its separate periods, each with its own ascribed date and span, and which graces the walls of geology departments around the world.

But he is also renowned for his Arctic geology. From the beginning of his career, he was drawn to the rocks of the remote Arctic, sure that he would find extraordinary geological secrets half-buried beneath the ice. He was right. By scouring the scarce rocks of the far North, Brian discovered the first traces of a global glaciation. He was the grandfather of the Snowball.

Brian's fieldwork, though, was never easy. In August 1949 he was leading an expedition over the ice fields of Svalbard, a frozen archipelago several hundred miles north of Norway and east of Greenland. He and his four companions had been away from base for days, dragging all their supplies with them. Now their route back led up a dauntingly steep slope of ice. If hauling on the flat is bad enough, uphill it can seem nearly impossible. Still, Brian had decreed that at the brow of the hill they could stop and camp; there would be food, warm drinks and rest. The five geologists duly buckled up and began the long, hard pull. Their heads were down, their attention fully focused on gaining the top of the slope. They had no idea of the disaster that was about to strike.

IN THE BEGINNING

* * *

THIS WAS Brian's second visit to Svalbard. Eleven years earlier, in 1938, he had been there as a young graduate, part of a brief student expedition. Svalbard's rocks had immediately intrigued him. They were among the oldest in the world, and many had formed in the Precambrian, that long Dark Age of the Earth. Brian realized that these rocks could provide a rare window into this ancient, mysterious time. But they were also remote and inaccessible, covered for the most part with a thick blanket of ice. In just a few places, dark, conical mountaintops and ridges of rock poked out above the snow. Brian had been intrigued by these outcrops. He'd caught glimpses of great rocky cliffs bearing giant folds and faults. Where did the folds come from? How had the mountains formed? What could they reveal about the workings of the world?

On the '38 expedition, there had been little chance to find out. The expedition had other priorities. Brian was the only geologist among a group of geographers. There were ice-forms to study, and maps to make, and not enough time for everything. And then the Second World War had intervened. But now Brian was back, thirty-two years old, a fully-fledged Cambridge academic running his own show. This time he could decide for himself where to go and what to study. The geology of the islands was a blank, and he was determined to fill it in. He wanted to understand every outcrop, every layer of Earth's prehistory.

This was Brian's first time as expedition leader, and he felt the responsibility keenly. He was a slight man with pinched features and a nervous disposition. Brian planned, some people said, to excess. Most of his time beforehand was spent worrying over details. For every problem he had a contingency. When it came to samples, notebooks, photographs, all the paraphernalia of a geological expedition, Brian's numbering systems were complex, consistent and legendary. Everything had its own alphabetical or numerical code. Every item of equipment slotted neatly and clearly into the overall plan.

The expedition food was chosen for high calorific content rather than taste. There was margarine, processed cheese, sugar, oats, biscuits, chocolate and a fatty mix of dried meat called pemmican, all the same items that had sustained Antarctic explorers like Scott and Shackleton just a few decades earlier. This simple, efficient and egalitarian diet had appealed strongly to Brian's utilitarian instincts on the '38 Svalbard expedition, and he saw no reason to change it. (In later years he would bow to the necessity of supplementing the dull basic rations with spices, delicacies and other extras, but he never really approved.)

He had, however, learned one important lesson. In 1938, everyone had been ravenous. The rations had been designed for Antarctic expeditions using dog teams. Man-hauling required much more energy than the food provided, and there had never been enough to eat. When you're constantly hungry, staying warm becomes more and more difficult. At night your dreams are laden with food. You fantasize about medieval feasts and sweetshops and huge, rich desserts. And when you wake, you have to force yourself to harness up to a heavily laden sledge while your stomach is gnawing and your limbs feel weak and tired. The food on Brian's '49 expedition might have been dull, but he made sure it was plentiful.

Brian's watchwords—a legacy, perhaps, of his Quaker upbringing—were fairness, order and efficiency. He had already set in place the rules that were to govern his expeditions for the next forty years. No hoarding of food was allowed. Rations were divided evenly, and your portion belonged to you until midnight. Anything you hadn't eaten by then reverted to the general pile. Also, you were strictly forbidden to bring any additional delicacies secreted in your pack. What one member of the expedition ate, everyone ate. You could break these food rules if you chose, but only furtively and with a guilty aftertaste. Few people tried. Brian was scrupulously honest, and his attitude somehow spread.[3]

He also judged people firmly by their dedication to the task in hand. To be part of his expeditions meant abandoning any perceived status or sense of entitlement. His was an Edwardian value system. Would you volunteer, were you willing, had you put in the necessary effort to prepare? (When I first met Brian, years later, I had to lay down all my academic credentials before he would speak to me. He wanted to know about my degree, my doctorate, how much research I had already done. When he was finally satisfied that I deserved his time, he was promptly generous with it.)

Brian believed that a person's work should speak for itself, and he abhorred the notion of pushing himself forward. Take the naming of geological features. In those early days of exploration in Svalbard, many researchers gaily named the places they discovered after themselves and their friends. To immortalize themselves, they chose magnificent mountains, giant rivers of ice, great macho structures. But although Brian would become the world's leading authority on Svalbard, you'll struggle to find his name on the maps. Eventually, you may spot one small smudge, close to the summit of the ice cap, bearing the name "Harlandisen". Brian's students think this is hilarious. An isen is a rather nondescript patch of ice, usually found between more interesting places. Even so, Brian is embarrassed by the accolade. Ask him how the name came about, and he will blush faintly and mumble that the Norwegians insisted.

Brian's students loved him. They followed his codes strictly and with loyalty. On his '49 expedition he had brought along eleven students from Cambridge, split into different groups for maximum efficiency. Several parties had already investigated the coastal regions, tooling along the fjord-ridden coast in sixteen-foot open whaleboats, which Brian had christened *Faith* and *Hope*. The rest had taken the largest boat, an eighteen-foot dory called *Charity*. ("It's a biblical reference," Brian says. "You know. 'Faith, hope and charity, and the greatest of these is charity.'") Brian had bought *Charity* for seventeen pounds. She was a marvellous boat, big, wide and solid as a rock, with space enough for a ton of equipment. She had borne the third party to an encampment at the tip of Billefjorden, in the northwest of the main island. From there, Brian led a small party of four students, the "Northern Survey", out into the unknown territory of Ny Friesland. The plan was simply to map the rocks and begin to understand what was out there. Though these rocks came directly from the time of the Snowball, Brian as yet knew little about them.

The first few days were good ones. In clear weather the party sledged and skied, measured angles, surveyed the landscape, made sketches and took carefully numbered photographs. All around them, the dark brown tips of mountains and rocky cliffs poked through skirts of ice. Snow dusted every dip in the rocks. And flooding down every gully and alongside every cliff were Svalbard's great glaciers.

Glaciers are giant bodies of ice, with a texture like a strange combination of rock and river. They are solid, like the ice cubes in a refrigerator, and form out of snow the way rock is made from soft mud or sand. If mud falls consistently down on to a seafloor, its grains will eventually squeeze together and solidify into rock. Snow does the same thing. Individual snow crystals are gorgeous works of six-sided filigree. But if they pile up over time, these crystals begin to amalgamate. They squeeze up against one another. Their delicate arms smash and break and weld together. They trap pockets of air, meld into a hoary substance called firn, and then gradually solidify into hard, white ice. And then the ice begins to move. Like water it flows downhill, but at a magisterial, glacial pace. Glaciers don't just fill valleys; they create them. Flowing ice may be slow but it's inexorable, and a glacier can carve through solid rock.

For polar travellers, glaciers make great highways; but they come with hazards, too. When ice flows, it splits into deep fissures and cracks. Snow then drapes these crevasses, hiding them from the unwary. Break through one of these snow bridges, and you will find yourself plunging into the cold blue heart of the glacier. Usually you can avoid this hazard, since snow bridges often reveal themselves as tell-tale dips in an otherwise smooth surface. Usually, but not always.

Around a week into the expedition, Brian's party was travelling down Harkerbreen Glacier to investigate the rock cliffs on either side of the ice when the good weather abruptly deserted them. Thick clouds descended all around, until they could scarcely see the way ahead. They sledged gamely on, but the weather hampered all their efforts to investigate rocks, and Brian began to feel nervous. With only two days' rations in hand, he decided to try a new route back. If they could reach the wide sweep of Vetaranen Glacier, to the east, the way would be easy even in cloud.

To be safe, Brian decided to scout out the route ahead. With him he took one of the students, Chris Brasher. Chris was just twenty years old, but he was a fit and accomplished mountaineer. (He was also a most talented athlete. Five years later he would be one of the two pacemakers who propelled Roger Bannister to the first sub-four-minute mile. Two years after that he would win his own glory with Olympic gold in the 3,000-metre steeplechase.) Leaving the other three behind, Brian and Chris found a tributary glacier that snaked upwards and eastwards toward Vetaranen. They climbed doggedly up the steep ice slope, always checking for the dips in the snow that marked the presence of a crevasse. But the surface seemed innocent.

Back with the rest of the team, Brian directed operations. The way ahead was worryingly steep, but once over the slope every-thing should be easier. They would take a sledge at a time, start-ing with the heavier of the two, the Nansen. Nansen sledges are wonderful inventions, still used by polar explorers today. Their wooden parts are lashed together with hide, making them lithe and flexible enough to snake over bumps in the ice. At twelve feet long, they're also a good protection against crevasses. Even if you break through a snow bridge, the sledge will usually span the gap and act as a safety anchor, allowing you to climb back out again.

The five geologists attached their harnesses to the heavily loaded Nansen and began to plod their way up the glacier. Step, heave. Step, heave. They had almost reached the top of the slope.

Then the ground vanished from under them.

The sledge and the nearest two people plummeted immedi-ately into a vast cavern of ice. One, two, three, the others followed, whipped backwards on their harnesses through a huge hole in the snow. The foremost man came last, his ski catching on the surface and tearing away from his foot as he fell.

Seconds later, all five found themselves miraculously alive, sprawled forty feet below the surface. Through bad fortune, they had broken through a wide, thick snow bridge, wide enough that the sled was no protection, and thick enough that it was invisible at the surface. But through good fortune, the bridge fell with them, so that all five had come to rest on a soft cushion of snow. And another piece of good luck: though the chasm continued down for hundreds of feet, the entire team and their sled had landed on a wide ledge of ice. There was only one casualty. Brian felt a pain in his right ankle and discovered that he couldn't stand on it. (He didn't want to claim any great injury. He later wrote that it seemed to be "slightly broken".[4])

Inside a crevasse, the temperature is many degrees colder than at the surface. Quickly your nose hairs and eyelashes are coated with a fine hoarfrost. Your face becomes numb, and begins to show white patches of incipient frostbite. The only light comes in feebly from the snow hole far above you, or as a blue gleam from the cavern's walls. For the next eight hours, Brian was forced to stay put in this ghostly glow while the four uninjured students began the rescue operation. First they crawled along the ledge until they found a place where it sloped up to the surface and a natural hole allowed them to climb back out. Back down the slope then, to where the second sledge held spares of everything, including ropes, a testament to Brian's meticulous contingency planning. Piece by piece, the students hauled every item of equipment up through the snow hole to the surface. They did their best to haul Brian out, too, but it proved impossible. Because the cavern's walls curved away from the dangling rope, Brian couldn't reach them to steady himself, and he swung and spun uncontrollably. Eventually he was lowered back down. He strapped on skis and shuffled slowly and painfully along the ledge to climb out the way the others had.

Outside, the cloud was still thick and low, heavy with the threat of snow. Brian and the team camped, on half rations, and considered their options. They were at least two days from their nearest food depot, and four or five days from base. The route they knew involved a steep downward slope and another long, heavy pull upwards again. But at least this way was definitely safe, and they resolved to take it. Broken ankle notwithstanding, Brian had no intention of being pulled along on the sledge. Each morning he would strap on his skis and begin a long, lonely shuffle over the snow. His right ankle was useless. He had to use his ski pole to point the ski in the right direction. Behind him the four students would finish their breakfast, pack the gear, and haul the sledges along in Brian's trail. Around midday they would catch him, and stop for lunch. Then they would continue on into the distance, leaving Brian to trudge painfully along in their tracks. By the time he arrived at the night camp, food was already prepared, tents were pitched, and he could fall into his sleeping bag.

After five days they finally reached base. *Charity* bore Brian back around the coast to Svalbard's main town, Longyearbyen, where he was ordered straight into the hospital. His broken ankle had finally earned him a "hot bath and excellent care", which, he later wrote, made him "the envy of the others" since they had to return to the privations of the field.

One of the other parties from the expedition, it turned out, had also fallen foul of the hazards of Arctic travel. As Brian later put it, *Hope* was all but lost. Boat and crew had to be rescued from mid-fjord by a rubber dinghy. But Brian had known all along that Svalbard wouldn't yield its secrets readily. His contingency plans had been effective. His students were eager for more. And the various field parties had only scratched the surface of the data to be had. By the time Brian reached Cambridge again, he was already planning his next trip. He insists still that he wasn't drawn by the romance of the place. What pulled him back to Svalbard, he says, were the *stories*. He wanted to understand what the rocks could tell him. He didn't yet know that the rocks of Svalbard held a more extraordinary secret than he'd ever imagined. Nor did he know what trouble that secret would cause him.

BRIAN HAD been pleased with much of the organization of that first venture, but during his next few expeditions to Svalbard he was continually testing possible improvements. He began to build up his equipment, buying a whole new set of Nansen sledges to distribute among different field parties. Even though the Nansen hadn't protected his party from the ice cavern, such wide snow bridges were rare, and in every other respect the sledges had been great. He even found a handy source of Nansens back in England—buying several from the film set of *Scott of the Antarctic*, the movie that was just about to catch Paul Hoffman's young imagination, across the Atlantic in Canada.

Brian also began to realize that self-sufficiency and self-reliance were the keys to operating in Svalbard. Anything he left to someone else carried the risk of failure. Materials that had to be shipped north every season could be lost in transit. Relying on someone else for transport by sea could mean hanging around for days by the quay. Gradually, Brian established a base for himself in Svalbard, where he could store goods over the winter. He set up mechanical and electrical shops there. He bought covered motorboats that could sail safely around the coast even in the choppiest of seas. His expeditions became like guerrilla raids. Every summer his geological parties swarmed over the ice of Svalbard. They set up survey stations and measured outcrops; they climbed cliffs, collected samples, and steadily filled in the blank map of Svalbard's geological history.

And as Brian returned to the islands time and again, one particular feature of the rocks began to bother him. In many of the outcrops that poked through Ny Friesland's sheath of ice, a strange red stripe stood out against the pale yellows and browns and greys around it. Up close, the stripe was a chaotic mix of reddish boulders and rocks, all shapes and sizes, bound together in a background of fine silt.

Brian had known this pattern all his life. He'd grown up in Scarborough, on the North Yorkshire coast of England, and as a child he would collect the alien stones that studded the cliffs there—the famous "boulder clays". The boulders embedded in Yorkshire's sea cliffs were an unmistakable signal that ice had been on the move. They had arrived fresh from Scandinavia, where glaciers had dragged them off the land and dumped them into the North Sea.

Glaciers don't just glide serenely over a surface; they grind into it. A glacier scratches and scours the bedrock with the boulders it drags along. It bulldozes yet more rocks ahead of its advancing ice front. The surface of a glacier can be littered with debris that has tumbled off the steep cliffs at the sides and is carried along with the slowly moving ice. Eventually the glacier will spill into the sea. Perhaps it will break off into chunks of iceberg that gradually melt and deliver their load of rock debris to the seafloor as individual "dropstones". Perhaps it will simply offload its rocks just a little way offshore. Glaciers dump similar rock jumbles on land, but land deposits tend to be eroded away by the action of wind and rain, and most of the really old ice rocks that have survived around the world were formed in the protected environment of a shallow sea.

That's exactly what had happened to create Brian Harland's mysterious red stripe, and he was baffled by it. Why should he be troubled by signs of glacier deposits in frigid Svalbard? Because he already knew that in Precambrian times, Svalbard was very much warmer than it is today. He was sure that when the rocks of ancient Svalbard formed, conditions there were positively tropical.

They must have been, he reasoned, because most of the rocks in the outcrops he was studying were tropical. They were carbonates, pale grey and yellow rocks made of the same stuff as seashells. These rocks, though, were born before shells even existed. Unlike the chalks and limestones formed in more recent times from the crushed shells of sea-creatures, these carbonate rocks had nothing to do with the presence of living things. Instead, they had formed from a purely chemical process in Precambrian seawater, and then rained down on to the seafloor to be compacted into rock.

And here's the important point: this process happens only in warm seas. Cold water clings to its carbonate; only warm water releases it. That's why carbonate platforms hold up the sunny islands of the Caribbean. You'll find them beneath the Great Barrier Reef, and throughout the islands of Indonesia. And you'll also find them on either side of Brian's icy red stripe.

What's more, in the carbonates below his red stripe, Brian found oolite, a strange type of rock made up of tiny spheres that are squashed together like petrified caviar. This bizarre texture is also utterly characteristic of tropical climates. Six hundred million years ago, the islands of Svalbard had clearly been hot. Finding signs of ice among oolitic carbonate rocks was bizarre, like watching a glacier march across Barbados.

Brian began to investigate further. How about northern Norway? There, too, he found Precambrian carbonates interrupted by a layer of ice rocks. Greenland? The same. Now he began to pore over published papers, marking out anywhere in the world where geologists had mapped ice rocks in the Precambrian, the mysterious geological period that was devoid of distinguishing fossils. They were everywhere. Every single continent had the clear traces of ancient ice.

And then a whisper started in Brian's head. Perhaps the ice was global. Perhaps it had been everywhere in the world. At first his main interest in this was an arcane geological one. If the ice really had once been everywhere, the ice rocks it had left behind could be a Precambrian global marker. This might be a way of matching time-slice to time-slice for rocks all around the world, shedding light on the Earth's otherwise obscure Dark Age.[5]

Then Brian realized something else, something much more important. This ice came just before one of the most dramatic periods in Earth's history: the great evolutionary explosion that created complex life. Perhaps the ice was the trigger. It might explain why the Earth moved from the Dark Age to the Age of Enlightenment. Brian knew that biologists couldn't explain this breakthrough. There were simply no theories that made sense. And he realized that he might now have the answer. A climate change as big as the one he was proposing—surely that would be enough to shake the Earth from its slimy idyll, and jump-start the true beginnings of biodiversity. Excited, Brian marshalled his arguments about this "Great Infra-Cambrian Glaciation" and set about writing them up.

BRIAN HARLAND wasn't the first person to propose a global ice age. A Swiss researcher named Louis Agassiz had done so nearly a hundred years earlier. Agassiz suggested that ice had run rampant around twenty thousand years ago, much more recently than the Precambrian. In this he was partly right—ice had indeed stretched beyond its polar bounds then. But Agassiz's ice age was nowhere near as extensive as he'd imagined.[6]

Agassiz came upon the idea of prehistoric ice in the 1830s, while studying the geology of the Alps. He surmised that parts of the Rhone valley had been carved out by ice that had long since melted, and found boulders transported far beyond the existing fringes of the Alpine glaciers. He then began to discover that other parts of the world also showed signs of extensive ice. Scotland, for instance, bore carved valleys similar to those in Switzerland, but no glaciers remained there. Putting the evidence together, Agassiz proposed that there had once been a mighty ice age with glaciers stretching from the North Pole down to the Mediterranean. Then he grew more ambitious. The ice, he declared, had been everywhere. He even claimed to have found glacial traces in the Amazon rain forest.

Agassiz was the first ice champion. In the end he succeeded in convincing a sceptical world that ice could *ever* stretch beyond its present polar bounds. Thanks largely to him, we now know that the polar caps wax and wane on timescales of a hundred thousand years or so. When they stretch to their largest size, the world enters an ice age. The most recent one, which finished just eleven thousand years ago, is also the most famous, the time of woolly mammoths, mastodons and sabre-toothed tigers. And the receding ice heralds an "interglacial", the warm time between ice ages that we are experiencing today. Though researchers still argue about what causes ice ages, most believe that they are driven by subtle changes in the amount of heat reaching the Earth as it wobbles in its orbit around the sun.

But Agassiz's ice ages were nowhere near as extreme as the Snowball. In the last of his ice ages, the sheet of ice that coated northern America reached only as far south as New York. Another ice sheet blanketed northern Britain and Scandinavia, but scarcely made it into mainland Europe, let alone as far south as the Mediterranean. Some pack ice spread outwards from Antarctica into the Southern Ocean, and New Zealand felt the chill, but that's as far as it went. Ice didn't get anywhere near the equator. Other than an occasional iceberg, most of the oceans were ice-free. Global white-out? It wasn't even close.

Agassiz had a religious reason for overblowing the extent of his ice age. He felt it provided concrete proof that a providential God intervened in Earth's processes. God, so Agassiz thought, had deliberately introduced the ice to wipe out all previous creatures and leave an empty and bountiful stage to be occupied by His chosen new race: mankind.

This part of Agassiz's theory tumbled in the 1850s, when fossil finds demonstrated without question that most species had survived the ice age, and that the ice couldn't possibly have been global. Agassiz's many critics had feared the implications of his dramatic white-out, and they were delighted by this development. Earth was still, after all, a well-mannered and temperate place. Its climate might oscillate over time, growing a little warmer at times, a little colder at others. But nothing too bad had really happened, nor anything too extreme.

In the hundred years that followed, several people had noticed the much more ancient ice rocks of the Precambrian. Geologists had mentioned them in passing. Sir Douglas Mawson, the renowned Antarctic explorer who knew a thing or two about ice, had spotted signs of them in South Australia, and knew that they could be seen around the world. "Verily," he said in an address to the Royal Geological Society of Australia in 1948, "glaciations of Precambrian time were probably the most severe of all in earth history; in fact the world must have experienced its greatest Ice-Age."

But nobody had run with it. After Agassiz's embarrassment, who would want to declare that ice could ever have been global? Who'd put forward such an extreme idea, and risk exposing himself to the inevitable ridicule? Someone, perhaps, who was happy to swim against the tide, who took things at face value, who was self-reliant, unconcerned about how other people viewed him and made his own rules about status.

Brian Harland had grown increasingly convinced that the Precambrian was a time of global ice, one that was vastly more dramatic than the recent puny ice ages. By 1963 he had prepared all his arguments and set off for an international conference in Newcastle, in the northeast of England. He was ready to put his idea before the world.[7] But he had, as it turned out, picked an unfortunate time to champion the ice rocks. They were about to go crashing out of fashion.

This was largely thanks to the efforts of John Crowell, a geologist at the University of California, Santa Barbara. John enjoyed

going to England. Though he lived in California, he had been sec-
onded to the Admiralty in London during the Second World War.
His background was in geology, but he'd retrained in meteorology for the war effort. He was one of the three scientists who predicted the height of the waves, both surf and swell, that would be
experienced by troops landing on the Normandy beaches. Now,
on his way to that same conference in Newcastle, he had a paper
to present that had arisen indirectly from those London experi-
ences.[8] He had accumulated evidence that most of the so-called
ice rocks around the world had nothing at all to do with ice.

Through his work in the Admiralty, John had become fascinated by the behaviour of the sea just beyond the shoreline. In
particular, he had started to investigate a set of undersea canyons
close to the California coast. Most geologists assumed that the
canyons must have formed on land, and then been flooded at
some point when the sea level rose. But John discovered that the
undersea world itself was a violent place, and that the canyons
had been carved out by massive mudslides.

The canyon floors contained a jumbled mess of rocks, sand
and stones that had been carried along on the back of the sliding
mud. John realized that this looked just like the supposed leav-
ings from a glacier. For him, the mixed-up rocks that had been
called "glacial" for decades were nothing more or less than the
effect of underwater mudslides. How to explain the ubiquity of
these rocks? Simple. You get mudslides everywhere.

John's idea caught on quickly. He'd written a few papers in
the late 1950s, and many researchers were assimilating his ideas.
Trends come and go in geology as in everything else. John's mudslides were the hot new thing—ice rocks, laughably old-fashioned.
When Brian tried to talk about his global ice in Newcastle, his

colleagues were scornful. Didn't he know about the new findings? Why was he still harping on about an interpretation that had been so clearly superseded?

On the bus back from the conference centre, Brian found himself sitting next to John Crowell. He told John about his glacial rocks in Svalbard, and about his idea of the Great Infra-Cambrian Glaciation. John's response was kindly, and almost unbearably infuriating. They're not really made by ice, John said. Tell you what. Why don't I get some funding to go to Svalbard? Then I can check out how your rocks really formed.

Brian ground his teeth. He knew his rocks. He also knew what he was doing. He didn't need someone else to go and check his work.

In truth, it's not too hard to distinguish ice rocks from those created by mudslides. When ice is on the move, the rocks that it carries are often scarred and scratched with lines all pointing in the same direction. You can tell from the types of rocks whether they have been transported from long distances, or originated close to shore. In your deposit you might find a boulder that appears alone and has gently distorted the lines of the mud around it. Such objects obviously fell on to the seafloor from a melting iceberg overhead. Brian knew what signs to look for, and he knew he was looking at the effects of ice. But nobody would believe him, and John Crowell's new theory stood.

Over the next few decades, Brian published more and more careful descriptions demonstrating that the Precambrian rock jumbles around the world had been created by ice.[9] John Crowell, meanwhile, travelled to all seven continents, scrutinizing the rocks, gradually ruling them in (though, ironically, he never went to Svalbard). In the end, John and the rest of the world conceded.

Brian, the man perpetually ahead of his time, turned out to have been right all along. "He was willing to take a flier," John says now, rather ruefully. "And he turned out to be more correct than us sceptics."

But that still wasn't enough for other geologists. There was more trouble for Brian on the horizon, in the form of an old theory: continental drift.

IN THE early 1960s, earth science was being shaken to the core. Before then, most people had believed that the continents were fixed in place. Afterwards, almost everyone believed that they shifted. Compared to this, any worries about the ice rocks seemed minor. For geologists, the safe, comfortable ground beneath their feet was suddenly moving. Everyone was talking about it.

Plate tectonics, as the theory became known, was the new manifestation of an old idea. Back in the early 1900s, the German meteorologist Alfred Wegener had already made the disturbing proposal that continents moved around the surface of the Earth.

Wegener was a man blessed with intense curiosity about the world around him. Though studying the atmosphere was his day job, he found it hard to resist almost any earthly mystery that came his way. The planet, Wegener felt, was teasing him with its secrets. Once, after a journey to the Arctic, he wrote about the marvel of the northern lights, and how tantalized he felt by them. "Above us . . . a powerful symphony of light played in deepest, most solemn silence above our heads, as if mocking our efforts: Come up here and investigate me! Tell me what I am!"[10]

Part-time astronomer, geologist, adventurer, he made many expeditions to the polar regions, and even dabbled in hot-air ballooning. (At the age of twenty-six, he set the world record along with his brother, by staying aloft continuously for fifty-two hours.) Though he made important discoveries about the physics of the atmosphere as well as the wayward behaviour of continents, he was repeatedly rejected for professorships of regular universities, mainly because he refused to confine his research to a single academic area.[11] He wanted to know everything at once.

Wegener came up with the notion of continental drift around Christmas of 1910. He was looking idly at a map of the world when he was struck by how snugly the coastlines of Africa and South America fitted together. They looked like two pieces of a jigsaw puzzle. He suddenly wondered whether they had once been part of the same continent, and had only later drifted apart. Intrigued, Wegener began to find other evidence around the world that disparate continents had once been connected. There were ancient floodplains of volcanic lava in Africa and South America, which matched like two halves of a coffee stain. There were animal fossils of exactly the same types and mixtures on both sides of the Atlantic. In place after place, the geology or the fossils matched uncannily between continents that were now far apart. He concluded that the continents must surely move.

The idea was bold, intriguing—and widely derided. Most of the geological world immediately rejected it. They had been taught from birth that the Earth's surface was safely fixed in place, and Wegener's alternative made them extremely uncomfortable. Wegener didn't help his case by coming up with a preposterous mechanism to explain how the continents moved; he mistakenly believed that they ploughed through the Earth's solid crust like an icebreaker, and at breakneck speed. It also didn't help that he was a meteorologist rather than a "proper" geologist, and yet was putting forward geological evidence to support his claims.[12]

But Wegener didn't give up. He pushed and harried and accumulated evidence, directing tireless energy and determination into trying to prove his point. But before he could convince the world that he was right, his curiosity finally killed him.

He died during an expedition to Greenland in 1930. The plan was to establish a station high on the summit of the ice cap, where a few researchers would spend the bitter Greenland winter. Through months of isolation and darkness, they would study everything: the wind, the weather, the stars, the auroras, the snow, the ice. But the expedition encountered problems from the beginning. Unyielding fields of pack ice stranded Wegener's ship off the coast of Greenland for an agonizing thirty-eight days. By the time he finally broke through to reach land, the summer was half over. Though he sent off his advance party to set up the central station, known as Mid-Ice, he was already worried that there would not be enough time to supply it fully for the long winter ahead.[13]

Eventually, but very late in the season, Wegener set out for Mid-Ice himself. He took a team of hired Greenlanders and fifteen dog-sledges weighed down with provisions. The conditions were appalling, and after one hundred miles of blizzard and intense cold, the hired hands rebelled. Wegener ploughed on with just two companions. By the time the three of them staggered into Mid-Ice, on 30 October, the temperature was fifty below zero, one of the party was badly frostbitten and they had no supplies left to deliver.

The situation was desperate. Mid-Ice had scarcely enough food and fuel to supply its two present occupants through the winter. There was no way that all three newcomers could stay as well. Wegener celebrated his fiftieth birthday on 31 October, and, taking Greenlander Rasmus Willumsen with him, he set back out again on to the ice.

The details of what happened next are sketchy, pieced together from the scant clues that Wegener left behind. Around 160 miles from base, it seems that he abandoned his own dog-sledge, and started to ski alongside Willumsen's. He always skied fast. "The journey must never come to a standstill," he had often told his companions on earlier expeditions. "The natural pace of the dog-sledges is the normal speed to which everyone else must adapt himself." Fine words for a young man. Not such a good idea for a fifty-year-old, in half-light and bitter cold, on a surface that had been whipped up into solid waves of ice by the driving wind. At some point during this frantic, lung-bursting dash for safety, Wegener suffered a heart attack and died. His body was found neatly buried in snow, the grave marked by a cross fashioned from his skis. Of his diary, and of Willumsen, no trace has ever been discovered.

Wegener knew the risks involved in his science. "Whatever may happen," he had written before the expedition, "the cause must not suffer. It is the sacred thing which binds us all together. It must be held aloft under all circumstances, however great the sacrifices may be. That is, if you like to call it so, my expeditionary religion. It guarantees, above all, expeditions without regrets."

The German government offered to send a battleship to retrieve Wegener's body for a state funeral. His wife refused, and so his body still lies somewhere deep in Greenland's ice cap.[14] One day perhaps it will find its way to the sea, encased in an iceberg. If so, when the ice eventually melts, Wegener's remains will be gently deposited like a geological dropstone on the seafloor.

* * *

WITH WEGENER'S death, continental drift lost its advocate and the idea foundered. But a few scientists still held the candle for him and his idea. And to anyone obsessed with both geology and the Arctic, Alfred Wegener was the perfect hero.

Brian Harland had certainly always loved continental drift, though he says that's because of the idea, not the man. The theory had been proposed in 1912, five years before Brian was born, and he first heard about it as a schoolboy. Delighted, he gave a talk about it to his school. By then, most geologists had discounted the theory, and his teachers were unimpressed. At Cambridge, continental drift brought Brian more trouble. Cambridge was one of the main centres of dissent about the idea, and whenever Brian mentioned it, nobody wanted to know.

But Brian wasn't particularly bothered. He believed in letting the facts tell their own story, and to him everything seemed to point towards continental drift. The eastern region of Greenland, for instance, had rocks that looked just like those in the east of Svalbard, several hundred miles away. But the ones in between, those of western Svalbard, were completely different, even though they were obviously from the same time period. Brian was sure that western Svalbard must have been a separate chunk of continent that had wandered north and thrust itself between the other two places. He was sure that Wegener was right.

Eventually, once again, everyone else caught up with Brian. Most of the world's rocks contain tiny magnetic particles of iron that act like compass needles, pointing towards magnetic north. Unlike compass needles, though, these particles can't swing around at will. When the rock hardens from its original soft sediment, they are frozen in place, and if the continents haven't moved since then, all the magnetic particles should still point north. But in the early 1960s, the tentative new science of rock magnetism began to reveal a surprise. Rocks on different continents had "compass needles" that pointed toward different "norths". The only explanation anyone could think of was that the continents had changed their geographical positions since the rocks first formed.

Then physicists began to get involved, and suddenly everyone seemed to have new evidence showing that the continents moved. Strictly speaking, the new theory of plate tectonics was different from Wegener's. Continental drift suggested that the continents themselves were moving. We now know that the Earth's entire surface is broken up into plates that shift around, some of them bearing continents on their backs. But still, in essence, Wegener's idea was vindicated.

It's ironic, then, that this vindication put yet another spoke into Brian's Snowball wheel. If the continents truly did move, then there was a much easier way to explain Brian's ice rocks than the outrageous idea of global ice. Everyone knew that the poles were cold and the equator was hot. So each continent must simply have drifted over to the polar regions to collect its ice, and then wandered away again.

Of course, Brian had already thought of that. He was a *champion* of continental drift. As he'd made clear in his papers about the ice rocks, he'd already tried to fit all the continents together in a huddle around the pole. But there was simply no way to do it. However he arranged his geological jigsaw, he couldn't cram all the continents into the polar regions. Some were always left out in the sun.

But plate tectonics was now on everyone's lips, and to many geologists, moving continents could explain everything. Brian realized that he had only one option. He had to prove that at least one of the continents was near the equator when the ice formed.

This, he figured, would be tantamount to proving that there had been a global freeze, since it's extremely hard to freeze the equator without freezing everything else, too. Our sun's rays come to us untrammelled, in single-minded parallel lines. At the equator, they strike the equator more or less full on. At the poles they always hit at an angle. Shine a torch directly on to a piece of paper, and you'll see a neat round circle. Now tilt the torch, and the circle will grow and distort into an ellipse—the same amount of light spread out over a greater area. The same thing happens on our spherical planet. Directly overhead at the equator, the sunlight is fiercely intense. But the further north or south you go, the more it spreads.

The upshot of our celestial geometry is this: it's easy to freeze the polar oceans and to make glaciers in Alaska and Antarctica, even at sea level where there's no thin mountain air to help. But the closer you get to the tropics, the harder it becomes to make ice. If the temperatures had somehow dropped low enough to freeze the equator, everything else must also have frozen.

To see if the equator really had been frozen, Brian decided to adopt the same technique that had been used to vindicate Wegener: rock magnetism. Many rocks come with their own magnetic birth certificate, because they adopt the local pattern of the Earth's magnetic field. This has a classic, characteristic shape. Stick one end of a piece of wire into the top of an orange, bend the wire over, and push the other end into the bottom of the orange. That will give you some idea of how the Earth's magnetic field looks. It shoots straight upwards at the poles and passes horizontally over the equator. If you were standing near magnetic north, the field would pass through your foot, say, and up out through your head. If you were standing near the equator, the field would pass through you horizontally, across your hips, waist and shoulders.

When they're young and soft, rocks are still impressionable; they can take on the stamp of the Earth's field. The magnetic particles they contain line up the way the Earth's field does. As the rock is compressed and hardens, these particles are fixed in place, and the direction they point in tells you where they were born. If their field is vertical, they were born near the poles. If horizontal, they come from the equator. The rock magnet is weak, to be sure, much weaker than one that you'd stick on your refrigerator. But it is just measurable.

Brian decided to try to find out if his rock samples from Svalbard and eastern Norway had fields that were horizontal. He built a new instrument, so sensitive that it could detect magnetic tremors from the elevator as it rose and fell in its shaft, fifty yards away. He measured sample after sample. At first he was thrilled. The Svalbard rocks showed a horizontal field—just what you'd expect if they'd formed near the equator.

But he couldn't really be sure. The weak magnetic field in the rocks might have been altered in the hundreds of millions of years since they'd been created. In the 1960s, rock magnetism was still in its infancy. There were no sophisticated techniques to rule this out. If the field had been altered after the rocks formed, there was no way that Brian would be able to tell.

Then came a devastating blow. The university authorities built a car park outside the lab. Any further magnetic experiments would be hopeless. Brian was already working at the limits of the available technology. The rock magnets were so weak, and the instruments to measure them still so crude, that any slight changes in the field around them would wreck the results. And now the magnetic field in Brian's lab changed every time a car entered or left.

He published his findings,[15] but he always knew that he hadn't made his case. Nevertheless, he continued to investigate the geology of Svalbard, organizing more than forty expeditions in all. Eventually he put his findings into a prize-winning book, the definitive geological guide to the archipelago, which contained 500,000 words and took him five years to write.[16]

Brian has never stopped working—he's not the retiring type. Even in 1990, at seventy-three, he was still studying Svalbard's rocks by day, and sleeping at night in a tent pitched on the shore. Thanks to their newly protected status, polar bears had grown bold by then, so the camp was surrounded by tripwire attached to a device that would fire blank cartridges to frighten off any marauders. But Brian was characteristically unfazed about this danger. The wire, he felt, was just a damned nuisance—set off many times by stumbling people, but never by bears. His final expedition was in 1992, but he has not stopped working on the Svalbard rocks in his collection. Now, at eighty-five, he still goes into his office in Cambridge every day.

Though Brian never succeeded in proving his Great Infra-Cambrian Glaciation, he was always obstinately convinced it was right. To bring the idea out of the cold, however, would require much more evidence. The next step would be to demonstrate without question that ice had been present at the equator. That would take an unusual scientist, someone with an eye for problems that were a long way out of the ordinary. Someone who was also a world expert in magnetism.

FOUR

MAGNETIC
MOMENTS

Joe Kirschvink adores magnets. You might say he's irresistibly drawn to them. In fact, that's just the sort of crummy joke that Joe would probably make. Even now, in his late forties, he loves to act the clown, with his bright button eyes, and brows that periodically shoot high into his forehead. He's compact of build, and fizzes with energy and ideas. His unruly hair and neat moustache are the colour of wet sand.

Joe is a professor at the California Institute of Technology, an august institution that lies among the villas of Pasadena in southern California. Caltech professors are hard-nosed people. It's one of the most fiercely competitive academic establishments in the world, filled with some of the most gifted scientists. They work long hours, know how to sell themselves, guard their patches jealously, and make sure they stay ahead. You don't often come across a Caltech professor like Joe, who constantly describes his own ideas as "nutty", and invites you to call him a nut. "Honestly," he says. "I don't mind."

In truth, Joe Kirschvink is one of Caltech's most brilliant brains. His strength lies in his ability to look at old problems in a new way. He delights in topics that other scientists shun, ones that have a whiff of the weird about them. Joe often does his work away from the scientific spotlight, but he tends to make the kind of discovery that swings the spotlight over to him. And then he moves on to something else. His motto could be "never dismiss, never assume". In his introductory geology class, he has each student write a "nut" paper, in which they have to consider an offbeat hypothesis, ideally one that has been ridiculed by the scientific establishment, and then describe how they would rigorously test the idea. His students love it.

Joe first began experimenting with magnets when his father inadvertently wrecked the family microwave with an exploding golf ball. (He had been told that warmer golf balls travelled further and was attempting to heat the ball up quickly.) Joe got the machine parts to play with. Later, as an undergraduate at Caltech, he used his genius with magnets to good effect in a tradition called "stacks". At one point in their final year, all the seniors are supposed to create elaborate locks for the doors to their rooms and challenge junior classmen to break in. Joe's particular one is legendary. From the outside, the door was blank—there were only a few magnets and some written clues. But on the inside of the door, Joe had rigged up a series of magnetic switches that had to be tripped in exactly the right order to open the door. Standing outside, the hopeful lock-breakers had to move a magnet to different points in the blank door using only the written clues as a guide. Each wrong move was punished with a loud blast of "The Ride of the Valkyries". The stack proved too inventive even for Caltech students. Nobody managed to break the lock and claim the two gallons of ice cream waiting inside.

Around the same time, in the mid-1970s, Joe began to experiment with naturally occurring magnets. While on a trip to Australia, he heard that "north-seeking" magnetic bacteria had been discovered in Massachusetts. These creatures had achieved a clever evolutionary trick. Bacteria usually find their food in the depths of ponds and puddles, so they have an incentive to know which way is down. In the northern hemisphere the Earth's magnetic field lines slant downwards, so for the bacteria "north" equals "down" equals "food". To find food, they simply clamp their internal magnets on to the Earth's slanting field lines and slide down, like firemen down a pole.

Joe was in Australia when he heard about these strange magnetic bacteria. He knew that in the southern hemisphere, the Earth's field lines pointed the opposite way, so that "north" there was "up". What, he wondered, happened to southern bacteria? He immediately rushed off to find likely looking grubby ponds and pools of water, and snared a few bacteria. Using the magnifying glass and magnet that he always carried with him, he found that Australian bacteria swam consistently south. Their inner magnets were upside down compared with those of their northern cousins.[1]

The Aussies loved it. Joe found himself unexpectedly on the front page of the *Canberra Times,* brandishing a beaker of bacterial sludge from the Fyshwick Sewage Works. He knocked the Ayatollah Khomeini off the headlines. His south-seeking bacteria became celebrities. Joe set them up at a Canberra University party so geologists could watch them swimming backwards and forwards as he flipped a magnet below them. One partygoer, peering over his shoulder, asked if they liked beer. Joe promptly applied a drop of Foster's lager to one side of the beaker, and then flipped the magnet to make that side "south". The bacteria galloped toward the spreading yellow liquid, but as soon as they tasted it, they turned tail. Australian bacteria apparently do not like beer. Later, Joe tested the northern bacteria in his hometown of Phoenix, Arizona. The American bacteria showed no inclination to turn tail at the Foster's-water interface. They swam directly into the beer, and promptly perished. "They died happy," says Joe. American bacteria, unlike Australian, have no idea when to call it a day.

Joe was just fooling around with the beer, but these experiments had an underlying seriousness. He became intrigued by the way the Earth's magnetic field affected the creatures on its surface. At Princeton as a graduate student, he discovered tiny, pure magnets in the brains of honeybees and pigeons and proved—to everyone's astonishment—that the creatures use these in-built magnets to navigate.[2] This was a classic case of finding a scientific basis for ideas that had previously been ridiculed. Pigeon fanciers had long believed they should not race their birds when there was a magnetic storm. Beekeepers were convinced their charges had an innate sense of direction. Nobody else believed them. Joe put science where the myths had been. His findings, made while he was still a student, are now described in even the most staid of textbooks.

Joe also found magnets in the brains of fish, whales and even humans.[3] Thanks to Joe, it's now known that we all carry tiny, built-in magnets around in our heads. These magnets may even help humans navigate, although Joe never managed to prove that we use our magnets the way bees and pigeons do.

Joe's obsession with magnets even extends to the names of his children. His wife, Atsuko, is Japanese, and his sons are called Jiseki, which means "magnetite" in Japanese, and Koseki, which means "mineral". Jiseki came first, in 1984. As the firstborn son, with a lineage that traced back through Atsuko to the Japanese imperial family, the child had to be given a distinctive and meaningful name—one that would be approved by the temple monk back in Japan. This approval hinged crucially on how the name looked when written down. One night, after many fruitless suggestions, Joe asked his wife what "magnetite" would look like. She wrote down Jiseki. The name had a curious shape, like two lightning bolts next to one stone over another. Atsuko called her mother in Japan, who immediately took the name to the temple. It passed every test that the monk set.

Koseki, "mineral", appeared two years later. Though his name didn't pass quite as many temple tests, it was still perfectly acceptable, and everyone agreed that it was the natural follow-up. And if there had been a third child? This is what Joe says about the name that would have been next. "As well as magnetite and minerals, I work on meteorites. The Japanese word for 'meteorite' is 'inseki'. Time to stop."

Now Joe lives in Pasadena, but Atsuko and the two boys live across the Pacific, in Japan. Joe goes there for three months every year. Atsuko was very unhappy in California, and Joe told me rather sadly that this was the only arrangement that seemed to work: "I've discovered that if you spend a hundred per cent of your time with someone, and they're only twenty-five per cent happy, then life is miserable. But if you spend twenty-five per cent of your time with them and they're a hundred per cent happy, life is much better."

Perhaps that's why Joe is so close to his students. He treats them as family. He gives them responsibility, but tempers this with endless support. They know he would go the limit for them, and they love him for it. They say he's "generous", "modest" and "brilliant". Everyone has a "Joe story" to tell. Here's a sample. Joe, they say, speaks two languages—"English" and "foreign". He travels the world doing fieldwork, and after every new field site, his "foreign" becomes more confused. Once, at a petrol station in Baja California, Joe wanted to thank the elderly Mexican man who had pumped his petrol and cleaned his windscreen. Joe had just returned from collecting samples in Russia and, confused for a moment about which branch of "foreign" he should be speaking, said, *"Spasibo"*, which is Russian for "thank you". The Mexican petrol attendant beamed with delight. *"Pozhalsta,"* he said. "You're welcome." This particular Mexican turned out to be a Russian immigrant. It could only happen to Joe.

All of Joe's research ideas involve magnets of some kind, and there's usually some kind of controversy involved. But those are the only connecting points. Wherever the magnets are, in biology, geology, chemistry, or astronomy, you'll find Joe. He's suggested a way that animals might use magnetic fields to sense imminent earthquakes. He's even worked on evidence for alien life in a meteorite from Mars. (This last work was done in an ultra-clean lab to avoid any earthly contamination. The sign outside says: "This is the Door to the Planet MARS. Only the purest in Heart, Mind and Body may enter here.")

Still, Joe doesn't seek out controversies for their own sake, and he's no contrarian. He just enjoys delving into areas where his unusual way of thinking can resolve disputes and mysteries. He's like a curious and energetic child, blown by the wind into following now this idea, now that, whatever catches his attention. There's a downside to this. Joe claims he does things for fun, not for recognition, and that's just as well. Not many fellowship and prize committees appreciate researchers who work in so many different fields and in such wildly imaginative ways. Joe's students whisper about this angrily, sure that he has been passed over for honours that they feel he deserves. But his critics would say that he spreads himself too thin; he deals in too many different areas, they say, and doesn't follow them through.

When the idea of a global freeze caught Joe Kirschvink's attention in the 1980s, he betrayed both his greatest strength and his greatest weakness. He made what's probably the biggest, most imaginative leap of anyone involved with the theory. Without his insights, the Snowball idea would almost certainly have foundered. But then came the weakness: he didn't follow through. He came up with the ideas, put together the picture, and then moved on to where the wind next blew him.

JOE'S PERSONAL Snowball saga started one day in 1986 when he received a manuscript from a prominent geological journal. The paper was on a topic right in Joe's area of magnetic expertise, and the journal wanted to know whether they should publish it. Not surprisingly, since the world of magnet fanatics is a small one, he knew the researchers who had written it. George Williams, then from the Broken Hill mining company near Adelaide, Australia, and Brian Embleton from Sydney, had been studying a small outcrop in South Australia. The outcrop was full of stones dropped by icebergs, and the other tell-tale signs of a deep freeze. And the researchers wanted to figure out where the ice rock had been born: near the poles, where you'd expect, or near the equator?

Remember that the magnetic field frozen into a wandering rock at birth tells you everything you need to know about its place of origin. If the field is vertical, the rock was born at the poles. If horizontal, then it came from the equator. And according to the paper that had appeared on Joe's desk, the ice rocks from South Australia had fields that were almost as flat as they come. This place, the researchers claimed, had been ice-covered within a few degrees of the equator.

Joe was unimpressed. This work suffered from the same limitations that Brian Harland's had. For the approach to work, the researchers needed to prove that the field was frozen into the rocks *at the moment of their birth,* something that Brian had never managed. There were several ways in which the field from the original homeland might have been subsequently wiped out and printed over. Heating rocks to high enough temperatures erases their magnetic memory—like leaving a credit card on a radiator. Water flowing through pores in the rock can also deposit new magnetic minerals there, which adopt whatever the field direction happens to be as they settle in place. For any of these reasons, Williams and Embleton could have been reading a fake birth certificate.

As Joe read carefully through the manuscript, he realized there were ways the necessary proof could be found. Times and techniques had moved on since Brian Harland's day, and there were now ways to tell whether a field had been reset. In the paper he was reading, Williams and Embleton hadn't included these crucial tests. Joe recommended rejection.[4]

But something about this manuscript had piqued Joe's ever-lively curiosity. Could there really have been ice at the equator? Joe thought he already knew of a cast-iron reason why the Earth simply couldn't freeze this way. Snow and ice are dazzling. They reflect sunlight. A shiny white Earth would send the sun's rays bouncing back into space. So if Earth ever got into that state, Joe thought, it should never be able to get out of it.

That much had been first suggested back in the 1960s. Just about the time when Brian Harland was investigating his Svalbard rocks, a young Russian climatologist, Mikhail Budyko, was playing with this idea: what would happen if you let ice run riot on the Earth? Budyko set up a simple model in which ice started off at the poles, but could grow as it wished, and then let the model run.

The result horrified him. White ice at his model's poles reflected sunlight, making the Earth a little colder. Because temperatures were colder, more ice grew, which reflected more sunlight, and so on. The ice in Budyko's model grew and spread and grew and spread until it became unstoppable. When white ice reached the tropics, it tipped over a threshold and the entire Earth froze.[5]

This was the "ice catastrophe". After it, there would surely be no way back. If the Earth had ever frozen over like this, its shiny white surface would have reflected sunlight back into space. That, Budyko felt, would be a disaster. The planet would cool catastrophically, and he thought that the ice could never melt again. Once you entered the ice catastrophe, he decided, there would be no way to escape. Earth would have been doomed to spin through space, frigid and lifeless.

Obviously that didn't happen. So, equally obviously, Budyko reasoned, Earth must never have frozen over completely. Budyko, and everyone else, concluded that his ice catastrophe must never have taken place. He seemed to have provided yet another reason why the Snowball could not have been.

Two decades later, in the 1980s, Joe Kirschvink knew all about Budyko's ice catastrophe. Everybody did. And he also knew its corollary: the Earth can't freeze. If modellers came up with a white Earth in their experiments, they simply threw those results away.

So what about this new evidence from Williams and Embleton? Perhaps, Joe felt, he should probe a little more. An Australian geologist whom Joe knew happened to be travelling to Williams and Embleton's particular part of South Australia at about that time, and Joe gave him a compass. "Pick me up a sample or two," Joe said. "Nothing fancy, just hand samples. But check their orientation when you chip them off the outcrop."

THE AUSTRALIAN outback has many guises. There's the famous dry red centre that houses Uluru and Alice Springs and the weird red mining town of Coober Pedy, the world's biggest producer of precious opals. The summer's heat is so fierce there that half of its meagre population lives below ground in mud "dug-outs", and its post-apocalyptic scenery is the darling of movie producers. But there's also a subtler wilderness, several hundred miles south of Coober Pedy, on the way back to Adelaide, and civilization. There lie the Flinders Ranges with their dusty valleys and row upon row of rounded mountains. They are softer than the red centre, their colours more muted. And they have had many parts to play in the Snowball story.

The main route to the Flinders winds through a narrow valley, at the foot of a sharply pointed mountain called Devil's Peak. Here, at Pichi Richi Pass, is where Joe's rock samples came from.

To get to the outcrop, you climb over a small wire fence on to a patch of bare, rocky ground scattered with tufts of dry grass. Winter's the season to go there—in a southern summer, the heat would be unbearable. Even in March, the temperature quickly climbs to 90 degrees F. or more. At least there is merciful shade to be had among a stand of eucalypts with their peeling bark and graceful bone white trunks, and the odd desert oak with its black spiny fruit shells still clinging to the bare branches.

Through the trees, a hillside slopes down to the floor of a dry gully. There are no trees on the hill, just patches of golden grass, and at its feet lies a jumble of mud-coloured rocks with a strange pink tinge. The rocks are stranger still up close. All of them are shot through with rhythmic dark lines, as if they had been painted in neat, careful parallels.

The lines, though, pre-date artists by hundreds of millions of years. They are the last remnants of ancient tides. Once this area was underwater, just offshore from an estuary. The tide flooded in to land, carrying with it a slurry of fine sand. As the tide ebbed, sand was washed back out to sea and deposited gently right at Pichi Richi. In and out, ebb and flow, as sand landed periodically on mud, the rhythmic patterns built up. In the end they solidified, becoming sets of regular dark lines in a pale reddish mudstone. Such rocks are called "tidal rhythmites", and they are very rare. A few good waves will destroy the pattern completely before it can harden into rock. But somehow the seafloor thereabouts was protected from waves, and the layers survived.

George Williams, the researcher whose paper had caught Joe's eye, had already used the rhythmites to work out exactly how long days lasted when the Earth was young. One day hasn't always passed in a little over twenty-four hours, as it does now.

Since the whirlwind days of its youth, the Earth has been steadily slowing on its axis and days have been getting longer. Those short early days are frozen into the rhythmic patterns of Pichi Richi's rocks. From the monthly tidal cycles that he measured there, Williams figured out the number of days in a Precambrian month, and the number of months in a year. When his rhythmites were still mud and sand, a little over 600 million years ago, a year lasted thirteen months, and a day less than twenty-two hours.[6]

But for the Snowball story, the rhythmites have a more important role to play: their neat inscribed lines show clearly where the rock slumped and folded as it was forming. Joe planned to use these folds for a definitive test of whether the rock's magnetic field came from its birthplace.

Imagine a slab of something flexible—an eraser, say. Now take a pen and draw horizontal lines along the side of the eraser, so it looks a bit like a layer cake seen from the side. If you bend the eraser into an arch, the lines will bend too, following the curve of the arch. However, if you first bend the eraser, and then draw lines horizontally across it, the lines won't follow the shape of the arch at all. They'll cut right through it. A fold test works the same way. If the magnetic field in the rock formed before it folded, the field lines will follow the curve. But if they were overprinted later in the rock's history, they will cut through the curve, ignoring its shape completely.

So Joe took a slab of rock from Pichi Richi whose dark tidal lines had clearly folded into an arch, and began to investigate. Did the lines follow the curve, showing they were original, or did they cut through it, showing they were overprints?

A graduate student made the painstaking measurements and then presented Joe with the results. Joe was astonished. The lines did seem to follow the curve. This changed the odds dramatically in his mind. Perhaps the flat equatorial field really had come from the same time as the ice.

And then he had another idea.

On a field trip to Canada, he had noticed that among the Precambrian rocks there were thick red layers of ironstones. That was mysterious. Ironstones belonged to a time much earlier in Earth's history, when oxygen first appeared in the atmosphere. Before there was any oxygen, the seas were full of dissolved iron that had come from the Earth's interior, pouring out of underwater volcanoes and deep-sea vents. But as soon as the air became oxygenated, the ocean's iron literally rusted. It turned into solid iron oxide and was sprinkled on to the ocean floor to become the layers of ironstone you can see in ancient rocks today. All that makes perfect sense. But the ironstones then stopped. Since the air was full of oxygen, dissolved iron could never build up in seawater the way it once had, and there were no more ironstone layers.

Until, that is, one bizarre iron blip a few billion years later, the blip that produced the ironstones Joe saw in Canada. They appear elsewhere in the world, and always in the same geological time period—just towards the end of the Precambrian, just before the first complex animals emerged from the slime, just around the time of the mysterious ice deposits. And then, shortly afterwards, they vanish again. The very late Precambrian is the only time other than that very early period when ironstones appear in the whole of Earth's history. The question is why.

Joe realized that he might now have the answer. Maybe the ice was the cause. If the oceans had frozen over, perhaps seawater had been cut off from the air long enough to accumulate lots of dissolved iron from underwater volcanoes. If the ice then melted and exposed all this iron to the air, then—boom!—it rusted again, and a new set of ironstones was born.

Everything was making sense. The magnetics from Pichi Richi seemed to show that there was ice at the equator. That's the hottest place on Earth. If the equator freezes, everything else has to freeze, too. And now the ironstones provided independent evidence that the oceans had frozen over at the same time. Four years before Paul went to Namibia for the first time, Joe was looking, suddenly, at total white-out.

He was thrilled, but also troubled. He wanted to believe in a global freeze, but what about the contrary evidence from Budyko's ice catastrophe? The Earth couldn't possibly have frozen over, or it would have remained frozen for ever. And yet Joe had seen the evidence for equatorial ice with his own eyes, and measured it with his own instruments. In science, as in life, when theory conflicts with evidence, it's usually the theory that's wrong. But try as he might, Joe couldn't think why. The ice began to haunt his dreams. He found himself tossing and turning at night, waking up in a sweat, thinking, "Did the Earth really *do* this?"

And then suddenly he had it. He conjured up a way out of the ice catastrophe. The melters of the Snowball, the evaders of the ice catastrophe, were volcanoes. Then as now, Earth was scattered with volcanoes, which periodically spilled out molten rock and heat. The Snowball wouldn't have stopped them. They could erupt perfectly happily, even under ice—as they do today in Iceland.

MAGNETIC MOMENTS

The lava from these volcanoes wouldn't itself have been enough to melt the Snowball. But when volcanoes disgorge their lava, gas comes too. Curling plumes of gas rise from the sides of an active volcano. An eruption can fling great clouds of gas high into the atmosphere. Gas bubbles up from hot vents beneath the sea. And one of the main gases to come from the heart of a volcano is also a villain in the world today: carbon dioxide.

Carbon dioxide, CO_2, is the gas that threatens us all with global warming. Every molecule of carbon dioxide traps a little heat. The more CO_2 you have in the sky, the more heat you trap. The effect is, famously, like that of a greenhouse. Carbon dioxide lets sunlight in, but prevents the Earth's body heat from escaping, providing a very effective and cosy way of warming a planet up.

And here's what Joe suddenly, thrillingly realized. Each volcanic eruption would pour a little more CO_2 into the sky, and gradually the greenhouse would heat up. Carbon dioxide would build up in the air, and wrap the Earth in a blanket of warmth. This blanket would trap more and more heat, and after millions of years the heat would finally melt the Snowball.

The idea is even cleverer than it sounds. In normal times, carbon dioxide doesn't usually build up in the air like that. Volcanoes are erupting all the time, but the Earth usually has a sort of built-in thermostat operated by rainwater, which strips out the excess CO_2. When rain falls through the air it picks up CO_2 and, as a result, becomes slightly acidic. The acidified rain lands on rocks and reacts with them chemically, handing over its load of dissolved CO_2. By this mechanism, any excess CO_2 is scrubbed out of the air and locked away inside a new rocky matrix. Over millions of years the Earth ends up more or less in balance, never too hot or too cold.

But if the planet freezes over and its rocks are blanketed with ice, the CO_2 thermostat switches off. Now there is all give and no take. Volcanoes keep giving out carbon dioxide, but the ice-covered rocks can no longer soak it back up again. Left unchecked in this way, the greenhouse effect of the CO_2 will build and build until it's ten times, even a hundred times, what we have today. With our puny little efforts to pollute our atmosphere by burning oil and coal, we'll maybe double the amount of carbon dioxide. Over millions of years the runaway volcanoes would have sent CO_2 levels spiralling beyond the wildest imaginings of any oil conglomerate. The Earth's atmosphere would have turned into a furnace.

The aftermath of the melting would have been hell on Earth. Dante, says Joe, would be proud of it. For tens of thousands of years—until the excess carbon dioxide was finally locked back up in the rocks, and the furnace switched off—any slime creatures that had survived the freeze would have found themselves scorched. These would be prime conditions, perhaps, to weed out all but the few, and to set the stage for the emergence of a whole new kind of life.

Joe's idea was ingenious. So what did he do with it? Embark on a round of conferences and lectures presenting it to his peers? Publish the idea in an acclaimed scientific journal? Not exactly. Though the academic world runs on conference presentations and published papers, Joe did little of either. The graduate student who made the measurements on the Pichi Richi fold presented the results at a single academic meeting,[7] but never wrote them up. Joe put together his thoughts about the Snowball in a tiny two-page paper, in which he gave his idea about the volcanoes just a few paltry sentences.[8] The paper took four years to come out in an obscure book, a vast monograph read only by the highly committed.

Joe still can't explain fully why he didn't grab his brilliant idea and run with it. Years later, when he read Paul Hoffman's work, he was chagrined. "Damn it. Why didn't I think of that?" But by then he had moved on to other things, tugged by his graduate students into their areas of interest. Perhaps he simply didn't have enough confidence in his deep freeze. It was one more "nutty" idea among the rest.

Still, Joe had made the strongest case to date that there was once ice at the equator. He had envisaged a worldwide freeze-over, and found a way out of the worst problem the idea faced. In a way, he had put the story together. What was needed now was evidence that he was right. Was there any sign left on Earth of his global super-greenhouse? Could anyone really show what had been happening to life in the oceans when the Snowball was in progress, and during its heated aftermath?

Meanwhile, Joe contributed another essential item to the story. Brian Harland had called this global freeze the Great Infra-Cambrian Glaciation. Joe was made of snappier stuff. He remembered as a child moving from Arizona to Seattle, where he endured three miserable freezing winters. At first the wet Seattle snow bemused him, but he quickly learned to pack it around a rock to make a snowball that had maximum impact. What was Harland's Infra-Cambrian Glaciation but snow packed around a rocky planet? Joe rechristened the glaciation with the name it has borne ever since: Snowball Earth.

He also did one more thing, something that would prove crucial. At a conference in Washington, D.C., in 1989, Joe found himself chatting with Paul Hoffman over dinner. Paul was still working in Canada at the time. He hadn't even encountered the Namibian ice rocks then. But as they ate, Joe cheerily told Paul about his latest crazy theory. That night he planted a seed in Paul's mind, and when Paul found crucial Snowball clues in the ice rocks of Namibia, the seed began to germinate.

FIVE

EUREKA

When Paul Hoffman became obsessed with the Namibian ice rocks, he sensed they would eventually reveal to him some important story about the history of the Earth. Throughout the 1990s, he returned every year to study them. He was where he wanted to be, driven by a new mission, as happy now in Africa as he'd ever been in the Canadian Arctic.

Each season he went eagerly from outcrop to outcrop, making every moment of his precious field time count. There was no such thing as a rest day. Any days not spent on the outcrops were for driving between them. Paul would often hike up and down hills, over rocks and down stream cuts and gullies. He went out fast and hard. He would work his way up through the sequence, noting what type of rock was there, how the rock formed, whether it came from a delta, a river, a deep ocean floor, how it related to the rocks above and below. He would put a sheet of clear Mylar over an aerial photograph and draw in the rock type with a sharp pencil. Sometimes he stopped to measure the rocks in a particular section. He took out a folding carpenter's ruler and climbed up the section, recording its thickness yard by yard, noting any peculiarities in his waterproof yellow notebook (waterproof to prevent the figures smudging from sweat—there's no rainfall in the Namibian desert during the dry season). He developed his own shorthand of neat hieroglyphs for sandstone and mudstone, diamictite and carbonate, and the sharpness of the contacts between them.

This was standard field geology. The first task in a new field site is always to build up a detailed picture of the whole geological terrain. Nothing geologists discover holds any weight amongst their colleagues unless they fully understand, and can describe, the context in which it is found.

Sometimes Paul collected samples. He would take a geological hammer and smash chunks off the rock face. There's something about holding a geological hammer that makes you want to hit rocks. Weigh one in your hand, and you'll find yourself itching to whack something with it. Still, to carve a hand sample into just the right shape for your pocket requires considerable skill. Paul is very, very good at it. If his geology career went awry, he could make a living as an ornamental rock chipper.

Perhaps he wants a sample of a particular structure in the rock, or something that shows exactly how the contact looks between two different rock types. He holds an unwieldy chunk of rock in one hand and chips away at it casually. Thwack, smack, smack, and all the useless bits miraculously fall off, all the right bits stay behind. The trick apparently lies in choosing the right sample, finding the flaws that will dictate where the rock breaks, and then hitting it at the right angle in the right place, with just the right amount of force. When there's anyone around and a sample to be had, Paul can't resist showing off his skills. It's infuriating to watch him. Also oddly inspiring. You want to rush off furtively and practise. You want to do it as well as he does.

Every day Paul would stay out on the outcrops as long as he could—too long sometimes. He often had to race back to camp before the heavy Namibian night fell. The surface of the rocks was rough, like sandpaper. If he tumbled on them in the darkness, his skin would be shredded. The nights were cold as well as dark. Paul's field season spanned the Namibian winter, when the days were still hot, but evening temperatures slipped quickly down into the thirties. Back in camp, the next race would be to build up the wood fire in a sandy hollow, and then to huddle around perched on the cooler boxes that serve as chairs.

Preparations for dinner take place by the light of a headlamp. Peeling and chopping is on a trestle table, spread with a garish plastic sheet. (And I mean garish. Think enormous pink and purple flowers, shocking even by torchlight.) There are usually fresh vegetables in the cooler, red peppers say, or beans, which can be a little frayed at the edges if it's been a couple of weeks since the last resupply. They go into a pot over the fire, along with onions and garlic and tins of fish, mussels perhaps, crabs, shrimps or tuna. And then there's rice or pasta or potatoes. Paul's wife, Erica, was astonished the one time she saw him cooking for himself in the Canadian Arctic. At home, Paul is determinedly hopeless at domesticity. Even in the field, the more culinary of his students wince sometimes at his vagaries. Paul puts cucumber into stews. Once a student trained in Italian cooking caught him adding ginger to spaghetti sauce. After a hard day in the field, though, you're ready to eat anything. You'll soak up the food with dark, heavy bread and sip Namibian beer, "the best in the world". And then, in the darkness, you rinse off your plate or bowl carefully, sharing the miserly dishwater allowance. Paul has his own utensils in Namibia, his own red plastic bowl and enormous coffee cup, white with a dark rim. He jokes about his possessiveness, but nobody else touches them.

Water is the really scarce commodity. Namibia has hundreds of miles of coastline, but not a single year-round river. There are plenty of river channels, and water can run in them briefly during the wet season, when it's too hot to work. But by the time Paul arrives in Namibia, all the rivers are dry. He has to carry his water with him in giant plastic barrels crammed into the back of the Toyota. Water sets the limit on how long he can stay in the field before he has to go into town for a resupply. It is reserved strictly for drinking and cooking. Washing is banned. You're even supposed to swallow the water that you use for brushing your teeth rather than waste it by spitting it out. The lack of water is a blessing, of sorts. Namibia contains plenty of dangerous wildlife, but most shun the dry regions; lions, cheetahs, rhinos and leopards all rely on open waterholes to survive.

Still, the desert has hazards of its own. Late one afternoon Paul was driving down the dried-out bed of the Ugab River. The light was starting to fade, and he flicked on the truck's headlights. He was starting to feel worried. In Namibia, darkness falls quickly, and it would soon be too late to find a decent campsite.

Paul wound his way hurriedly down the canyon on the sandy river floor, dodging the rocks and branches swept there by an old flash flood. Up ahead his lights picked out a thick black log, maybe nine feet long, lying in the sand. The Toyota could handle that, no problem. But at the last minute Paul swerved around it, striking what might have been a glancing blow. The log had seemed to twitch as he passed.

He was intrigued. He slowly backed up, craning his neck to see the scene illuminated by his white tail-lights. The log had vanished. No, it was standing up, and heading towards the vehicle, fast. It was chest high, four and a half feet above the ground, just about the height of the Toyota's open window. Now Paul could see that it had curious yellow rings the length of its body. It was a zebra snake, a western barred spitting cobra. It had spread its black hood angrily around its face and it loomed unnervingly large in the wing mirror. Paul remembers wanting to laugh. This was like the *T. rex* scene from *Jurassic Park*. "Objects in the mirror are closer than they appear."

But he also knew that zebra snakes were deadly. The toxin would quickly paralyse his muscles, and shut down his breathing. He had no serum as an antidote, since serum has to be kept cool and Paul had no refrigerator. Without immediate artificial respiration, he would suffocate. If someone pumped his lungs with their own air constantly while he was rushed back along these twisting canyons, in the dark, out along the bush tracks to the nearest village and then on and on to a town that perhaps had a hospital, he might survive without too much brain damage. Zebra snakes don't even need to bite you. They are called spitting cobras for a reason. Normally they are excessively shy, but when aroused they can spit their cytotoxic venom six feet or more. This one was clearly aroused, and Paul hastily rolled up his window.

The snake book in the passenger door of Paul's Toyota contains many lurid pictures. Alongside the featured snakes from southern Africa, you can see the human effects of their venom:

rotten arms, legs and hands, attached to bodies with pained, hopeless faces; limbs and torsos with puncture points surrounded by skin that is black, blue, yellow, swollen, pitted and blotched. "Don't read the snake book," Paul says to every newcomer, to first-time field workers and naïve young graduate students. "It will only give you nightmares." Everybody immediately opens the book and stares.

You are told, when you first come to Namibia, never to unroll your sleeping bag until the very last minute, just before you climb in. Each morning, when you wriggle out of the bag, you immediately bind it into a tight bundle. Everybody knows about the sleeping bag left unrolled at Khorixas rest camp by an unwary student, about the zebra snake that slid inside during the day and was there waiting for him when he retired to his tent. He survived, just, since he was relatively close to town. When you're camping out in the remoter parts of the Namibian desert, you don't need to hear this story twice.

Paul refuses to be worried by the Namibian snakes. He says that they're shy, rare and usually more than happy to avoid him. In all his years working in Namibia, he's encountered only one other serious snake—a black mamba, the most aggressive and deadly of all the ones in Paul's gruesome book. Even then, Paul didn't see it himself. And the geologist who did disturb it from its rock barely had time to gasp before the snake had vanished.

Thanks to the lack of open water, there's little else to worry about. Except, that is, for the desert elephants. They can dig for water. By using their tusks to create wells and waterholes, they can survive in some of the slightly less parched parts of the desert—the Huab River, for instance, towards the Namibian coast, where water is relatively easy to come by. Though the riverbed is dry, ground water lies not far below, and the surroundings are unusually verdant. Pungent African lilies poke up through the sand, surrounded by unexpected pockets of green. There are spiky euphorbia bushes and stands of mopane trees and twisted acacias.

The elephants in the Huab needn't be a problem. Paul has often camped there. Unlike, say, Canadian bears, you can easily discourage elephants from visiting camp. Though the riverbed, which makes a perfect campsite, is also a highway for elephants, they prefer to take the shortest, easiest route on their nocturnal journeys. Pitch camp on the wide part of a bend, and they will generally leave you alone. Even when they occasionally do wander around a field camp at night, they tend to be respectful and disturb nothing. The next morning, you simply wake to see their giant prints in the sand—ridged ovals the size of serving platters, or snowshoes, XXL. Paul isn't bothered by elephants any more than he is by snakes, or than he was by the flies and bears in Canada. Though they're dangerous, he says, elephants also tend to be shy. But they get angry when aroused. When I went to the Huab, I quickly found this out.

THE TUSKS were the first thing I saw—short, white and wicked. Then the rest of the elephant's head took shape against a backdrop of dusty acacia leaves. Tiny eyes set in a creased, anxious forehead. Ears thrust outwards, making the great head seem monstrous. This, I dimly recalled, was the elephants' universal warning signal. "Ears back: good. Ears forward: very bad." I froze.

African elephants are immense creatures—they weigh up to six tons and stand some eleven feet tall. They're fabulous, viewed from the window of a safari truck. But between me and this tusker lay just fifty yards of bare, scuffed sand, flanked with thorn bushes. I had found no trace of my companions in two hours of hard hiking. The camp was further still—direction unknown. I was lost and alone. I was also—now—in big trouble.

This predicament was partly my fault, but not entirely. Paul Hoffman had invited me to see his field site in Namibia, and I'd been counting on him to see me safely through. A few hours earlier he had laid out the plan for the afternoon. Five of our party were to squeeze into a vehicle and negotiate the wide sweep of the Huab River's dried-out bed, while the remaining four of us hiked to meet them across a sandy basin, carpeted with thorn scrub and occasional groves of acacia trees. We all knew where we were heading. Paul had pointed out a rich red outcrop of rock, standing against the skyline a few valleys away.

So I was frustrated but not unduly alarmed when I realized that—characteristically and without warning—Paul had charged off into the bush to begin the hike some minutes before, without checking to see who was following. His two harassed field assistants, more accustomed than I was to this habit of his, had apparently grabbed their packs and plunged after him. Hastily, I picked up my camera and ran, calling as I went. No sign. Back I went to the vehicle to check the final destination. But the vehicle had gone.

Of course, I could have walked back to the camp—a mere thirty minutes or so away—and spent the afternoon in craven contemplation. But I could see the outcrop quite clearly from where I stood. Nothing was stirring—even the air was still. I couldn't resist. This was my chance to show Paul how well I could manage alone in the bush. (Why did I want to? I don't really know. Paul has that effect on people.)

The day was beautiful. Though the rains were long gone, there were still green patches of grass, crowned with a silver-gold sheen where the tips had begun to dry. Between the grass and twisted black thorn trees were bare patches of sand spattered with mustard-coloured mosses. I even felt a frisson of delight when I came upon old elephant dung—cannonballs of dried grass and mud that marked a network of tracks winding through the tough, thorny scrub of the river valley. "Elephants make the best paths," Paul had told us, and it's true that the trails they blaze are easy to follow—wide, sandy and obstacle-free. My chances of spotting an elephant were remote, but at least I could use their tracks, taking my bearings from the jagged outcrop of rocks on its distant hillside.

After more than an hour of hard walking, I finally reached the dry Huab riverbed. There on the sand were two sets of vehicle tracks, but both looked old even to my inexperienced eye. There was no sign of human footprints. The elephant track cut diagonally across the river, heading in the direction of the outcrop. I followed it.

I was hot and tired by now, and increasingly discouraged. But then, in the distance, I saw a distinct brown shape—an elephant—loping across the sand. Enchanted, I stared as it crossed the river and began to climb the far slope. I followed as closely as I dared, watching in delight as it found shade under a large tree and began lazily twitching its ears. Then, planning how I'd boast about my sighting to the rest of the crew, I marched on.

My new enthusiasm didn't last. The outcrop was getting no closer and I had still seen no human signs. Relief when I heard shouts of "Oi! Oi!" evaporated when I realized they weren't human voices but those of baboons, barking warnings from up on the hillside. In the distance I saw another elephant crossing the river, trunk trailing in the sand. Now my close-up on wildlife seemed much less enchanting. The air was cooler, and the afternoon was drawing in. Suddenly everything was stirring. With growing unease, I waited until the elephant was out of sight, and then continued cautiously along the riverbed.

Then I heard the roar. The quintessential lion sound. The noise you'd make at the zoo to tease kids. I reasoned with myself. Lions are rare in the Huab. Unlike the elephants, they need standing water to survive, and this place is far too dry. They're common in Etosha National Park, far away to the northeast, but I'd be very unlucky to find one here. Perhaps I just imagined it.

Right on cue, the roar came again, from the dense patch of scrub directly ahead. This scrub lay right beside the escarpment that sloped up to the outcrop, and to find my companions I had to walk past it. I squared my shoulders and strode ahead, pinning everything on finding the van, finding my team, finding something safe. Along the escarpment, around the side of the outcrop, not running, not smiling, I squinted up into the sunlight, searching for signs of human life. There was no one there.

I panicked.

Suddenly all directions looked equally alien. A civilian in geologists' territory, I'd foolishly kept all my attention on the outcrop up ahead, and taken scant notice of the landscape through which I was passing. Looking back now, I could see no landmarks that I recognized. Blindly I plunged into the bush, and the thorn branches tore at me as I fought my way past. A full ten minutes passed before I forced myself to stop and try to think clearly. I had to find the camp before daylight failed. But how?

Hansel and Gretel. Though I'd left no white stones as markers, all I had to do, I realized, was follow my footsteps back. Here in the bush I'd trod mainly on springy grass, and there were almost no prints to follow. But I could head back to the escarpment, and up till then I'd mostly walked on sand. I resolved to find my footprints and follow them back exactly the way I came, no guesses, no shortcuts. And I began to calm down. At the foot of the escarpment I found the first clear footprints. More deep breaths, and I headed back along the elephant track.

There was no warning. Suddenly the large, angry elephant appeared, blocking my path, scarcely fifty yards ahead. Ears outstanding, it clearly wanted me out of the way. Still, I had a mad impulse to take a photograph. I resisted. These beasts scare easily. Three months earlier a Namibian man had been trampled to death just north of here, when he surprised an elephant and then tried to run for it. Climbing trees doesn't help. Make no threatening gestures. Slowly and carefully get out of the way.

Behind me was no go—that's where the elephant was heading. And I had no desire to get closer. I turned to the side and walked out into the riverbed. The great head turned too, and watched, thoughtfully, ears still spread, as I slowly, steadily, began to cross the river. In the open sand I felt even more vulnerable. If it charged, what then? Don't think, just walk.

I reached the other side, turned, headed homewards. The elephant hesitated, and then it too continued on its way. Stopped, stared and started again, this time rubbernecking. That mighty creature and I walked past each other, on opposite sides of the river, heads turned, each watching the other's every move. I have no idea how long I had been walking before I was finally clear. But by the time I heard human shouts and found the real outcrop—the one that I'd passed inadvertently, hours earlier—I was exhausted. Now that the fear had gone, I was furious. Paul knew I was a neophyte who had no idea where we were heading. How could he have abandoned me like that? How could he be so self-absorbed?

Paul greeted me with a cheerful smile. There was still time to see the rocks, he said, but I was in no mood to admire them right then. Why had he left without checking that I was there? Why hadn't he stopped when he realized I wasn't with his group? Confront Paul at your peril. Criticize him head-on, and his temper will flare. How dare I, he responded. It was my responsibility, not his. I should have let him know I was following him. (But this had been his plan, not mine. And who else would I have been following? How else would I have reached the outcrop?)

I stalked off to look at the rocks. They were beautiful, a delicate rose colour, and as I watched the sinking sun spill on to them, I tried to calm myself down. What's the point of letting it get to you? You know what Paul's like. He dishes this out to everyone. It's nothing personal. By the time I returned, Paul was wreathed in smiles again. He congratulated me on having acquired the day's best campfire story and offered to show me, as we walked back, the baboon skull he'd found on the way. He was charming and I was soothed. And I'd had my first taste of the strangely mixed experience of working with Paul Hoffman.

PAUL SAW the ice rocks everywhere he went in Namibia. He sought them out, and they intrigued and confused him. The rocks had formed in a Precambrian shallow sea, and though each outcrop was different, all bore the distinctive signs of ancient ice. Some contained lone boulders that had been dropped by icebergs floating overhead. Some contained the mad jumble of rocks and stones that had been scraped off the nearby land by glaciers, and bulldozed into the sea. Occasionally these jumbled rocks bore scrape marks where the ice had dragged them over the ground.

Some deposits were hundreds of metres thick, while others were just a thin skin. Many were also capped by a mysterious layer of carbonate rock, which had often turned pink, or perhaps ochre, with the touch of wind and weather. Paul, like Brian Harland before him, was baffled by this. Carbonates usually show up in warm water, in the tropics, but these appeared immediately after ice. And the contact between the glacial rocks and the carbonates was always knife-sharp, as if there had been some sudden, dramatic change from ice to tropics, from cold to hot.

After the summers mapping the Namibian ice rocks, Paul spent his winters puzzling over them back home. At the end of each season he brought samples to Harvard, and stared at the rocks in his office. He crushed them and measured them in his lab, all the time wondering what story they had to tell. Then one day, he remembered that odd conversation he had had with Joe Kirschvink years earlier about his "nutty" Snowball Earth idea.

Off to the library Paul went, to look up Joe's work on the Snowball. There was little enough, just that one short paper buried in a vast, obscure book. Paul read the paper and was gripped. He quickly dug out Brian Harland's research, and the magnetic work by George Williams, and Mikhail Budyko's papers about the ice catastrophe. He couldn't get enough of the Snowball. This was a story indeed. But each of the proponents of the Snowball had dropped it, one by one, starved by lack of evidence. What if Paul could provide the evidence? What if his Namibian rocks held the clues to this extraordinary catastrophe?

Now Paul started looking directly for Snowball evidence, not in the ice rocks themselves, but in the carbonates that bracketed them below and above, geologically before and after. Geologists have many possible tools for extracting stories from stones, and one of the best involves measuring the ratio of their isotopes—heavy and light versions of the elements they contain. Carbonate rocks, for instance, contain different isotopes of carbon. There's a lightweight version called carbon-12, and a heavier one called carbon-13. Comparing the ratio of the two can usually tell you something about the seawater that the rocks were formed in. And when Paul looked at the lab results from his Namibia samples, he was astonished. They had a bizarre carbon isotope signature, one that he had never seen before, with much less carbon-13 than he had anticipated.

Usually, ocean water and the carbonates that it produces are both rich in carbon-13. The ratio gets skewed to the heavier side because of the activity of living creatures. Bacteria in the ocean need carbon to grow, and carbon-12 is their favourite flavour. They grab carbon-12, and leave carbon-13 behind. Think of a box of red and green jelly beans. As you gradually pick out the red ones, the rest of the box will start to look more and more green. The same thing happens with carbon isotopes in seawater. When bacteria grow and grab carbon-12, the seawater ends up with proportionately more carbon-13. This seawater carbon is then bound up in carbonate rock.

So, when life is flourishing, the carbonate rocks formed at the same time have the skewed heavy seawater ratio. That's what was so strange about Paul's rocks. For carbonates, they were extraordinarily light. Before the Snowball, and for what looked like a long time afterwards, life apparently wasn't active at all.

Paul felt this was important, but he couldn't figure out what it meant. He was more baffled still by the "cap" carbonates that came after the ice. These are the same rocks that show up all around the world. Brian Harland had seen them in Svalbard. They stretch for miles in Australia, Canada, almost everywhere that the glacial rocks appear. And that's peculiar. One of the first things you learn in geology is that the Earth is emphatically not one big layer cake. Sure, individual regions might end up with layers of different rock types, cut through by rivers the way a knife cuts through a cake. But the rock layers are still different in different places. Take a snapshot of the rocks forming today on Earth, and here's what you might see. One place might have sandy seafloors or beaches that eventually solidify to produce sandstone. Somewhere else might be in the act of producing mudstone. Perhaps some volcanoes spew out their lava to cover another region with black basaltic rocks, and elsewhere you might find rocks that have been pummelled and transformed by the inner churnings of a mountain belt. The Earth is a very big, very patchy place. You simply don't get single events that blanket the entire planet with one type of rock. Period.

So where did these cap carbonates come from? Everywhere the ice rocks appeared, the caps seemed to be. And the ice rocks showed up on every single continent. Why? *Why?*

The caps also contained strange textures. The strangest were in the rock outcrops around the Huab River. There, set in a fawn-coloured cliff of carbonate, Paul found brown tubes resembling burn marks made by long, thin pokers. From a distance they were dark vertical stripes marring the cliff face for hundreds of feet. They weren't just on the surface of the cliff, either. Where chunks of rock had broken off, you could see more tubes marching on into the interior. In some places the breaks had sliced horizontally through the tubes, exposing them as a neat array of dark circles, each the size of a penny. The tubes looked like the regimented burrows of a highly organized worm colony, but there were no worms in the Precambrian. Paul was baffled by them.

Across the valley from the tubes, a student of Paul's found something just as strange. He had climbed up a steep ridge to inspect the carbonate outcrop. At the top was a plethora of huge rose-coloured crystals. They stood out against the pale carbonate rock around them, looking like giant splayed paw prints set into the vertical rock face. Or like the kind of feathered fans that Victorian ladies carried to the opera, though some of them were as tall as Paul himself. When Paul first saw these fans, he was astonished. He thought at first that they were fossils of some kind. They looked almost like giant clamshells. But no clams existed in the Precambrian, nor did any other creature that could make shells like these. The fans had to come from some bizarre physical process. But what?

Crystal fans, tubes, ice rocks, strange isotopes. Paul was increasingly convinced that all this evidence added up in some way, and would somehow yield crucial clues about the Snowball. He tried continually to make sense of all these features. He visited and revisited the strange carbonate formations to collect samples, to note and map and muse.

By the end of 1997, Paul was feeling frustrated. He had now spent five years doing fieldwork in Namibia, and he still had no research papers to show for it. He finally decided to write a paper about the carbon isotopes, even though he didn't yet fully understand them. Over Christmas and on into January, he perfected his paper, which was destined for a small journal.[1] He talked about the glacial rocks, and the strange isotopes in the carbonates that bracketed them. He talked about, and discounted, several possible explanations for the ubiquitous ice. And then—right at the very end—he suggested that Joe Kirschvink's ideas about the Snowball might provide a possible explanation. For once in his life he was being cautious. Not by choice, though. He was truly wrestling with the Snowball conundrum.

Until, that is, another new player entered the scene—a young colleague of Paul's, Dan Schrag. At the time, Dan knew nothing about the Snowball idea. He knew nothing about Brian Harland's work, or Joe Kirschvink's. He knew nothing about Precambrian rocks or Namibian geology. But there was one thing he knew plenty about, and that was carbonates.

DAN SCHRAG is Paul's best friend at Harvard. They look almost like partners in some comedy routine. Paul is in his sixties, tall, thin, shock-headed and white-bearded. Dan is in his thirties, short, plump and blond, with thin hair and a smooth round face. Dan's manner is smooth, too. He is supremely sociable. People are his thing—his networks stretch through every scientific field. He has hordes of friends. Many of them are hot young scientists like him, but there's a smattering of other types, too—artists, designers, people he knew at school. Every year Dan rents a house somewhere beautiful with five particular college friends. Spouses and children come, too. When a child is born to one of the group, there is a complex gifting scheme. Each group member provides the child with three favourite books to be read now, and two hundred dollars to be used later. The money is invested in a college fund, earmarked for entertainment purposes only.

Dan doesn't take anyone with him to these college reunions. He hasn't met the right girl yet. Instead he throws himself into his work, with sharp eyes and a quick wit and the veneer of arrogance that often comes with high intelligence. Friends say he is warm and generous; enemies say he is calculating. Everyone says he is brilliant. Dan has just won a MacArthur Foundation "genius" award: half a million dollars to spend as he wishes. He has decided to use the money for building a science retreat near the ocean on Cape Cod. The house will have skylights, an inglenook, a huge kitchen, plenty of rooms where Dan and his many friends can gather, cook, talk science and think.

Princeton gave Dan a professorship when he was only twenty-seven. He moved to Harvard four years later and was quickly, precociously, given tenure. Normally you'd work up to a place like Harvard. Normally it would take years to get tenure there. You'd expect to be in your forties, maybe, to have lots of research years behind you. Not Dan. He'd already published seminal papers by the time he hit thirty-four. By then, he had more ideas than he knew what to do with.

Dan loves ideas. He loves bouncing them around, tasting them, testing them, and seeing how they might work. He loves having intense conversations about them. Especially with Paul. Late at night, before he leaves the Geological Sciences Department and heads for home, Dan calls in on Paul. Even at eleven o'clock, or midnight, Paul is invariably still in his office. Dan sends his dog, Max, on ahead. Max, an amiable black Akita, knows the way. He ambles through Paul's outer lab, turns right and wanders into the office, walks up to where Paul is sitting at his desk, and thrusts his nose into Paul's hand. Dan stands outside and listens. He can tell Paul's mood by the tone of his grunt.

What follows is usually intense. Dan often ends up staying in Paul's office bouncing ideas back and forth for one hour, two hours . . . until the early hours of the morning. Sometimes the discussions become heated, but Dan isn't afraid of Paul. Heated arguments don't bother him much. "Everyone who's ever worked with Paul really closely has had a catastrophic falling-out with him," Dan says. "I know they're all watching me, waiting for the hatchet to fall. And yes, Paul and I fight. I don't just mean those tongue-lashings—I get those ten times a day. I mean violent, stand-up, screaming fights. 'You are the scum of the earth!' But I don't think I am the scum of the earth, so that's OK. Many's the time I've vowed never to speak to Paul again. But I keep coming back for more, because I like him."

"You know why Dan and I work so well together?" Paul often says. "Because friction creates heat."

Paul and Dan depend on each other. Though Dan is a professor in his own right, he still loves getting Paul's attention. He offers his ideas to Paul the way an eager puppy might. Once, when I was walking down the street with Dan at a conference in Edinburgh, we spotted Paul up ahead. It was late in the evening, and Paul was strolling hand-in-hand with Erica. Dan raced to catch up with them. As we all stood rather awkwardly outside Paul's hotel, Dan poured out his latest idea in a torrent of words, excited, watching all the while for Paul's response. Something to do with how many kilotons of carbon are bound up in Paul's carbonate rocks. Erica turned to me, looking amused. "How many kilotons of this have *you* had?" she murmured.

And Dan is Paul's conduit to the outside world. He's a Paul antidote. He deals with people smoothly and easily, soothing the feathers that Paul invariably ruffles. In some ways their relationship is uneven. Dan has plenty of people he can talk to about his ideas, but Paul doesn't have many friends. Perhaps because of this, Paul was at least as nervous as Dan about the tenure process. Paul was safe at Harvard. He was tenured. He could stay as long as he liked. But if Dan hadn't been awarded tenure, Paul would have lost his closest friend.

Over the months that Dan was being considered for tenure, Paul grew increasingly nervous. He gave evidence to the committee in Dan's favour. He tried to gauge how the decision might go, and spent most of decision day pacing up and down the department's corridors. Four minutes after Dan received the congratulatory phone call, Paul appeared in his office and collapsed on the couch in an exhausted, relieved heap.

Dan was the perfect person to ask about the Snowball rocks. The key to the story surely lay in those strange carbonates with their fans and tubes and weird structures. The ones that had somehow formed in the Snowball's aftermath, when the ice had finally gone. Carbonates come from tropical oceans, and Dan is an expert in them. He is an oceans-and-carbonates man.

Unlike Paul, Dan doesn't collect his samples from dusty deserts. Instead he tips himself off the back of a boat, usually in some gorgeous tropical location. The tropics are the Earth's heat engine, Dan says, because the sun's rays fall most intensely there. They are the key to understanding the Earth's climate.

Finding climate records from the tropics isn't easy. In polar regions the ice caps are like time capsules. Every year snow falls, and with it come dust and other chemical clues to the climate *du jour*. Gradually these layers build up, are buried, and turn into ice. When researchers drill down into the ice, they can uncover a climate record going back for millennia. In Antarctica, for instance, the ice cap is so thick that the layers at its base are more than 400,000 years old. There are bubbles of ancient air in the ice to be measured, too. Want to know precisely what the prehistoric atmosphere was like? Look no further. Here in this bubble is a whiff of air that was last breathed by *Homo erectus,* and was trapped and buried long before the Neanderthals ever appeared.

In temperate zones there's not much ice to be had. But researchers can at least study the thickness of tree rings. They don't even have to chop the trees down. They carefully bore into the side of the tree and pull out a thin core of wood, about the size of a drinking straw. And then they just need to count and measure. A thick ring? That was a wet summer. A thin one? Must have been dry. If they pick big enough, old enough trees, researchers can build up a climate record going back hundreds of years.

Dan, however, takes a different approach. His obsession is the tropics, where ice caps occur only on high mountains, and trees don't tend to grow annual rings because one season is pretty much like another. Instead he seeks out the climate records hidden inside giant corals. Like trees, corals lay down a new growth ring every year, but their rings are made of carbonate rock rather than wood. Measure the growth rings year by year, and you can learn how the climate of the tropical ocean has changed. If the coral is big enough and old enough, its rings can take you back hundreds of years. Find an ancient, fossilized coral, and you might even learn about the tropical climate from thousands of years ago.

When Dan dives for samples, he carries along his scuba gear, lift bags and a huge, one-hundred-pound drill. Even underwater, that's heavy. You dive in pairs. One person kneels on the coral with the drill; the other holds on to the rest of the equipment. The noise from the drill is so overpowering that you can hardly hear yourself breathe. When you're drilling, the fish keep well away. A fine powder gradually emerges from the edges of the drill hole, floats over the coral surface, and then disappears into the surrounding water. You stop every so often to add an extra length of tube above the drill bit. You wear thick knee pads to keep your wetsuit from being cut to pieces by the coral's spiky surface. Sometimes there are strong currents. You have to wear extra weight belts and try to anchor yourself on to the coral, and the sensation is like trying to drill the road during a hurricane. Sometimes the coral is deep, say sixty feet or so below the water, and you have only about an hour of air to land on the coral, drill a core, heave the drill up again, and race back to the surface. But if the coral is shallower you can take your time, enjoy the scenery, sneak some moments when you've finished drilling to wedge yourself in among the corals and sponges and watch the fish go by.[2]

Though Dan had never worked on rocks as old as Paul's, he had spent plenty of time studying corals and unlocking their carbonate secrets. And his insights into Paul's mysterious Namibian carbonates were about to prove crucial.

HARVARD, SUNDAY, 15 FEBRUARY 1998

PAUL COULD still feel a slight ache in his legs as he sat at his desk in Harvard. The Boston Marathon was coming up in a couple of months, and he had upped his training levels. Yesterday he had run almost twenty-two miles. Now, though, he was back in his office. He had worked through the evening and it was getting late, but he still didn't feel like going home.

That's when Dan wandered in, with a friend who was giving a seminar at Harvard the next day. "Come and meet Paul," Dan had said, over dinner. On a Sunday night? At this time? Oh yes, no problem. Paul will be there.

Dan wanted to talk to Paul about his Namibia paper. Paul had handed it to him a couple of days earlier. He'd scanned it, and felt annoyed because the interesting part seemed to be buried right at the end. As soon as Paul asked him about the paper, Dan waded in. You want to know what I think? This stuff about the Snowball is fascinating! You can't bury it! You can't just put one little sentence in about the implications. You need to think about this idea more. What does it *mean*?

Paul didn't need a second invitation. He had been hoping Dan would bend his brain to the Snowball story. So Paul told Dan about the strange cap-carbonate rocks and about how unexpectedly light the carbon isotopes were. He told him about the weird textures: the huge, graceful crystal fans, and the wormlike tubes.

He also told him the whole story as it stood then. That the Earth had frozen over, top to toe, pole to pole. That during this Snowball, volcanoes had continued to spew their greenhouse gases over the frigid Earth. That over millions of years the planet's atmosphere had become scorching. That this super-greenhouse catastrophically melted the ice. That the greenhouse gases stayed around afterwards for tens of thousands of years, blasting the Earth into a hothouse until they finally subsided.

Dan listened carefully. He retreated into a corner to think, while Paul and the friend politely chatted. When Dan is concentrating on a problem, he goes silent. His eyes dart around, focusing on something out of view. He often bites his bottom lip. Then, when he comes upon an answer, his eyes fire up. He immediately, eagerly, blurts it out. Wait! I've got it! I can explain the carbonates!

His idea was brilliant.

The planet was in stasis, cryogenically preserved. A thick layer of ice covered the oceans. Great glaciers crept and ground their way over the rocky surface of the continents, slowly pulverizing everything in their path. Ice bred more ice as the shining white surface repulsed sunlight, locking Earth in the mother of all winters.

So it was, and so it would always have been, but for volcanoes that poked above the ice or squatted on the seafloor. They erupted, as they always have, and each eruption spewed out ash and lava and—above all—carbon dioxide gas. Gradually, slowly, this volcanic gas built up in the air, wrapping the Earth in a blanket of warmth. And in the end, fire conquered ice. Drip, drip, came the first sounds of change, then trickle, then flow, then flood. Then meltdown. The ice vanished, and Earth went from icehouse to hothouse in a geological instant. So much was Joe Kirschvink's vision.

But now comes the new part, the part that suddenly dawned on Dan one Sunday night in Harvard. That hothouse, he realized, was like the tropical heat engine gone mad. The ice had gone, but the heat that melted it remained behind, on full blast. Dry, scorched air sucked up moisture from the oceans and whirled it into storm clouds. Hyper-hurricanes raced around the Earth's surface, flinging their watery burden back on the ground in torrents. And that burden was no longer just water. The air was filled with carbon dioxide. Whatever rain passed through it turned immediately to acid.

What did the acid rain fall on? An inviting layer of powder. Over millions of years, glaciers had ground the continents' rocks into dust. Ground-up material is always easier to react with. Think how much faster sugar dissolves when it's not bound up in a lump. In the post-Snowball world, that combination of ground-up rock and torrential acid rain was a chemical factory waiting to happen. Rock dust and acid met, mated and were swept off into the sea. They set the waters fizzing and foaming, creating a Coca-Cola ocean.

And then a new snowstorm began, this time underwater. All around the world, the post-Snowball ocean turned milky with flakes of white. They poured down on to every inch of the ocean floor. From the chemistry of acid rain and rock dust had come a massive outpouring of carbonate, which blanketed the entire planet. The flakes squeezed together, and hardened and turned into rock. They were the cap carbonates. This was Dan's idea. The cap carbonates, he said, arose directly from the intense, bizarre conditions that had rescued the Earth from its Snowball.

Dan and Paul both pounced on the idea and began to probe it. Did it work? Could it explain Paul's other conundrums? First, the strange tube rocks and rose-coloured crystal fans. Both could have come directly from the ocean's effervescent fury. The tubes might have formed when bubbles of gas shot upward inside the fast-forming carbonates. The crystal fans might also be some weird by-product of this frantic fizzing. In acidic hot springs like the ones at Yellowstone, you often find fan-shaped crystals, their arms radiating outwards with the sheer pace of precipitation.

Next, the rapidity of the change. The contact between carbonate and glacial rocks was always knife-sharp. That's just what you'd expect if the carbonate formed immediately after the ice melted.

What about the isotopes? Remember that the rocks showed a light, "lifeless" signal both before the Snowball and then for a long time afterwards. To explain what happened before was easy. Living things in the oceans pick out the light carbon atoms, the "red jelly beans", and leave the heavier carbon behind for the carbonates.

Paul had already suggested that before the Snowball, life's pace was probably slowing down as a reaction to the growing ice. Fewer living things meant less pickiness, and more light carbon left around to be bound up in the carbonates. That, Paul felt, was why the carbonates grew steadily lighter as the ice approached.

But afterwards was trickier. Life must have rebounded quickly after the ice melted, but the light, lifeless signal continued on in the carbonates for tens of thousands of years. Perhaps the "light" signal in the aftermath of the Snowball had nothing to do with whether or not life was flourishing. This intense formation of carbonate rock would swamp any normal signal. Carbonates were madly precipitating everywhere. They would be grabbing so much of the ocean's carbon, both light and heavy, that it wouldn't matter any more how picky the bacteria were being. It's as if you were sedately choosing red jelly beans from the pile, when a greedy cousin came along and snatched handfuls of the lot, both green and red, faster than you could eat any of them yourself. The post-Snowball carbonates were light because they swamped the signal from the bacteria. It all made brilliant sense.

Back, forth, back, forth. Whatever piece of evidence Paul could think of, Dan managed to fit neatly into his scheme. When they assumed a Snowball, all the pieces fell into place. The carbonates and the isotopes weren't mysteries any more. They were just what you'd expect. They were *predictions* of this new, improved Snowball theory.

The two of them grew more animated and excited. Try it this way and that. Look from every possible angle. The more they probed, the more Dan's idea really did seem to tie everything together. Brian's ice rocks, Joe's volcanoes, Paul's isotopes and cap carbonates, all added up into one elegant story. It was intoxicating. This, at last, was what Paul had been seeking. He could scarcely believe his luck.

Dan didn't leave until nearly three. After he had gone, Paul sat in his office, staring at his computer screen. At 3:04 A.M. he sent Dan an e-mail. The subject line was "funk in deep freeze" (the name of a jazz album). The message said, "Muchas gracias for tonight. I needed it badly. Thanks for the kick in the ass."

The next day, Dan was back. He'd been thinking all night about the Snowball. He wanted to work with Paul on a new paper. And he wanted to send it to *Science,* one of the world's most prestigious and highest-profile journals. This was to be their first direct collaboration, and Paul was delighted. For the next few weeks Paul and Dan wrote and rewrote and discussed and argued. They haunted each other's offices. They fired ideas at each other, stopping wherever they happened to meet. Students going to classes often had to step over them as they sat in the stairwells of the Geology Department, thrashing out the latest details of their theory. Here's something I just thought of! Hey, I've just made another connection! They both describe this period as the most exciting of their lives.

In science, good luck can be as important as good judgement. When Brian Harland first came up with the Snowball idea, he was too far ahead of his time. But Paul Hoffman became obsessed by the ice rocks in Namibia at the perfect moment. The Snowball stage was set when he and Dan experienced their eureka, and the key criticisms that had long dogged the Snowball idea were already solved. There was no ice there? Virtually everyone now believed Brian's argument that the rocks were formed from the action of grinding glaciers, and dropped by overhead icebergs. There was no way out of the ice catastrophe? With his super-greenhouse, Joe

Kirschvink had found a way that the Earth could go into a deep freeze and still recover. All the continents had been huddled around the frigid poles at the time the ice appeared? Thanks to the magnetic records from the Flinders Ranges, everyone now believed that in at least one place, in what is now the South Australian outback, there was ice within a few degrees of the equator.

For Paul and Dan, it could hardly have been more perfect. Their predecessors had each, individually, solved the arguments *against* the theory. Now they themselves had produced evidence *for* the theory. Paul's isotopes were evidence that large numbers of living things perished before the Snowball, and Dan's carbonates were evidence for the super-greenhouse that came after the ice.

Coming up with the Snowball story—understanding how the new evidence fitted in with the work that had already been done—took a particular combination of capabilities. Paul had the deep knowledge of Precambrian geology, the long years of field-work in Canada and Namibia. Dan had the understanding of how oceans work.

Telling the story would require yet another important combination, but this time of personalities rather than knowledge. Paul had the vehemence, the stubbornness, the single-minded obsession. Dan had the social network, the grace, the names and phone numbers of smart, imaginative scientists in many different disciplines. Unlike Brian Harland, Joe Kirschvink, or any of the other people who had worked on the glacial rocks in the past fifty years, Paul and Dan were prepared to run with the Snowball idea. Having put the theory together, they wanted to taste it, test it and spread it around. Paul in particular. This idea felt different from any others he'd been involved in. This was his chance, finally, to make a world-class difference to the way we all understand the Earth.

SIX

ON THE ROAD

Paul and Dan's Snowball Earth paper was published in *Science*.[1] There was an immediate flurry of interest, and the next step was to turn that flurry into a storm. Dan began calling his friends, and Paul took to the road. In the autumn of that year Paul went from one institution to another, purveying the good news. He was a very talented speaker, giving persuasive lectures that were both clear and impassioned. He had never, he said repeatedly, been so convinced that something was right.

Science is often messy. When you're judging any new theory, it's rarely as simple as yes or no, right or wrong. This is particularly true in geology. Reading the messages hidden in rocks is a craft, and different researchers invariably notice different things.

Theories in geology can rarely be accepted or dismissed out of hand. Even the ones that turn out to be broadly right often need to be massaged and modified, and given the initial benefit of the maybe.

But when a big new idea hits the scene, there's almost always a pattern of polarization. Though a few researchers keep a genuinely open mind, others immediately entrench either into pros or cons. These vehement souls will fight, criticize, and try to pull one another down. The survival of an idea can depend as critically on the quality of the rhetoric as on the robustness of the data.

That's exactly what happened when Paul went on the road. The more he promoted his idea, the more other researchers reacted against it. Many of them did so *because* Paul was promoting his idea so vehemently. He made no secret of his fervour. What he did, relentlessly, was force his opponents to face the Snowball theory, in a continual stream of public lectures and private seminars, papers, comments, reviews, e-mails and faxes, on stairwells at conferences, over lunch and around the campfire. He worked to get influential scientists on board and—yes—to squash those who disagreed. Sometimes it seemed as if he was trying to achieve the Snowball Revolution by the sheer force of his energy.

He even divided the science world into Snowball "believers" and "non-believers". When he was accused by one bitter critic of founding the "Church of the Latter-day Snowballers", Paul found the comment merely amusing—partly, I think, because he had a great riposte: "Someone once asked Charlie Parker if he was religious and he said, 'Yes, I'm a devout musician.' Well, it's the same for me. My approach to geology is that I'm a religious fanatic."

And Paul's relentless advocacy of the Snowball brought him continual criticisms that he lacked that most precious of scientific commodities: objectivity. His response was to liken himself constantly to one of his heroes. Alfred Wegener, the German meteorologist who had first championed continental drift, and who had perished at the age of fifty on the Greenland ice cap.

Paul saw plenty of parallels between his Snowball idea and Wegener's theory. Like the Snowball, continental drift (or, more strictly speaking, its later incarnation as the theory of plate tectonics) can explain many disparate puzzles under one elegant umbrella. As soon as you allow the continents to move, many other things follow. Where continents separate, they produce oceans. Where they collide, they make mountains. Where plates rub against each other, they can stick and suddenly slip, rumpling the Earth's skin with an earthquake. New seafloor is formed along the hitherto mysterious great ridges that run through the centres of the oceans like giant backbones. Old seafloor disappears by plunging down trenches at the edges of continents. Volcanoes form in the crust above these trenches. When the wet seafloor plummets into the Earth's interior, water creeps up into the overlying rocks, encouraging them to melt and spill their hot load on to the planet's surface. One idea explains all.

Also like Paul, Wegener was vilified as much for his *approach* as for his ideas. Wegener offended his opponents by the very way he reported his research. In his book *The Origin of Continents and Oceans,* he described his initial insight as an "intuitive leap". Intuition, many geologists felt, had no place in science. And there was worse to come. When Wegener discovered that fossils from South America uncannily matched those found in Africa and that the geology matched eerily too, he performed, he said, a "hasty analysis of the results of research in this direction in the spheres of geology and palaeontology, whereby such important confirmations were yielded that I was convinced of the fundamental correctness of my idea."[2]

He made a "hasty analysis"? He was convinced that he was correct? These sorts of comments deeply disturbed geologists, who felt that Wegener was far from dispassionate about his ideas. "My principal objection to the Wegener hypothesis," thundered one critic, "rests on the author's method. This, in my opinion, is not scientific, but takes the familiar course of an initial idea, a selective search through the literature for corroborative evidence, ignoring most of the facts that are opposed to the idea, and ending in a state of auto-intoxication in which the subjective idea comes to be considered objective fact."[3]

Another critic, Bailey Willis, said that Wegener's book describing the theory of continental drift gave the impression of having been "written by an advocate rather than an impartial investigator". (Willis wrote a paper[4] about Wegener's theory in 1944, which he titled "Continental Drift, Ein Märchen" [a fairy tale].) Joseph Singewald claimed that Wegener had "set out to prove the theory . . . rather than to test it" and accused him of "dogmatism", "overgeneralizing", and "special pleading".[5]

Ever since Wegener's vindication, proponents of a controversial idea in geology like to align themselves with him, and Paul does this a lot. Of course, there was nothing to prove conclusively in 1912 that Wegener was right, and he might have turned out in the end to be wrong. But still, the Wegener story is a cautionary tale to all geologists not to dismiss extraordinary ideas out of hand.

"Good ideas, when they're young, they're vulnerable," Paul said to me at a conference in Reno. "They're a pain in the ass, so you want to trash them. But the danger is the old ideas that everyone has got comfortable with. With a new idea, you have to cultivate it and let it grow and see where it takes you, and if you do, I think you'll learn faster where it's wrong than if you stomp all over it."

A few weeks later he sent me the following quote from Mott T. Greene, a biographer of Wegener:

Throughout the entire course of the debate [about Wegener's theory] neither his supporters nor his detractors seemed to have the clear grasp of a theory which comes from having read it carefully. The reason for this is a kind of guilty secret: most scientists read as little as they can get away with anyway, and they do not like new *theories* [Greene's emphasis] in particular. New theories are hard work, and they are dangerous—it is dangerous to support them (might be wrong) and dangerous to oppose them (might be right). The best course is to ignore them until forced to face them. Even then, respect for the brevity of life and professional caution lead most scientists to wait until someone they trust, admire, or fear supports or opposes the theory. Then they get two for one—they can come out for or against without having to actually read it, and can do so in a crowd either way. This, in a nutshell, is how the plate-tectonics "revolution" took place."[6]

Paul was clearly convinced this was true of the Snowball. He was certainly doing his best to force other geologists to face up to the theory. But it wasn't just Paul's *style* that made his colleagues object to the Snowball theory. There was another reason why many people found the theory discomfiting and even dangerous: it was a theory that required its proponents to think what to geologists was unthinkable. The Snowball Earth was different in almost every characteristic from the planet we see today. Accept Paul and Dan's theory, and you have to imagine our home planet behaving like Mars or Europa or some other alien place. That was more than enough to make many geologists shiver.

The problem was that the Snowball—as Paul described it—violated a key geological maxim called "uniformitarianism". This rule was first articulated in the eighteenth century, when geology in its modern scientific sense was born, and all geologists learn it at their mother's knee. It says that the present is the key to the past. The general assumption behind this rule is that the same things happening in the world today have been happening throughout Earth's history. Uniformitarianism is generally a good rule of thumb.[7] Try to explain baffling evidence from the past by invoking changes in the way the world worked, and you risk straying into the world of mysticism and magic rather than accessible, empirical science.

But there are some phenomena that don't show up in the everyday world, and yet are no less scientifically valid for all that. Uniformitarianism encountered one of its most serious challenges in the 1980s, when Walter and Luis Alvarez succeeded in convincing most—if not quite all—of the scientific world that the dinosaurs were killed when Earth was struck by a giant asteroid.[8] Their theory was, at first, most unpopular. Invoking some outside celestial agency for the dinosaur extinctions contravened the law of uniformity; it was like attributing it to an act of God rather than to some ordinary and perfectly explicable Earthly process. But then researchers found a huge crater from exactly the right time, off the coast of modern Mexico. Just because you can't see mighty asteroids hitting the Earth and destroying untold species of animals and plants today, the message ran, that doesn't mean it never happened.

And there are plenty of other reasons not to trust a simple reading of the world around you. You could do very sensible and careful experiments in the everyday world and end up thinking that time flows smoothly, that rulers are the same length for everyone, and that clocks tick at the same rate regardless of where you are or how fast you're moving. All of these assumptions are wrong. On large enough or small enough scales, the world doesn't work like that at all. Clocks can tick more slowly or quickly, time comes in packets, objects can be in two places at once, and the faster something is travelling, the shorter it gets.

All of these things were discovered in the early part of the last century, when relativity theory and quantum physics shattered our comfortable connections between direct experience and natural laws. There's a reason our intuition is often wrong: we evolved that way. In our normal lives we don't deal with relativistic or quantum scales. Nor do we deal with vast geological timescales.

And that fact has not escaped Paul and Dan. They both have a habit of talking about the Snowball as an "outrageous hypothesis". This is a nice touch, with an instant resonance for all geologists. The phrase comes originally from William Morris Davis, who, like Dan and Paul after him, was a professor of geology at Harvard. In 1926, inspired by the extraordinary happenings in physics at the turn of the century, he wrote a famous paper entitled "The value of outrageous geological hypotheses". Here's what he said:

> Are we not in danger of reaching a stage of theoretical stagnation, similar to that of physics a generation ago,

when its whole realm appeared to have been explored? We shall be indeed fortunate if geology is so marvelously enlarged in the next thirty years as physics has been in the last thirty. But to make such progress, violence must be done to many of our accepted principles. And it is here that the value of outrageous hypotheses, of which I wish to speak, appears. For inasmuch as the great advances in physics in recent years and as the great advances of geology in the past have been made by outraging in one way or another a body of preconceived opinions, we may be pretty sure that the advances yet to be made in geology will be at first regarded as outrages upon the accumulated convictions of to-day, which we are too prone to regard as geologically sacred.[9]

Davis *wanted* people to take risks in geology. And he was sure that any important new theories that stood a chance of invigorating the study of rocks would be outrageous. They would fly in the face of our intuition, just as the new theories of physics had changed everyone's assumptions about how clocks and rulers behaved.

That poses a particular problem, because geology is as much art as science. After geologists have painstakingly assembled all the evidence that rocks have to offer, they still need a certain amount of intuition for the interpretation. With physics or chemistry, you can test different mechanisms one by one. But geologists are fond of saying that their experiment has already been done. They can't rerun the Earth with slightly different conditions and see what happens. Instead they often have to use their instincts.

And Paul and Dan are convinced that when it comes to the

Snowball, you can't trust your instincts. This was a world that didn't obey normal rules. "The Snowball is a different planet," Dan says repeatedly. "You can't judge it by the same criteria we use today." Instead you have to trust the evidence, however strange it appears to be. And you have to be able to interpret it by thinking out of your skin.

This is something that suits Paul very well. He has spent his life running counter to convention. His passion for music began as a teenager when he became gripped by atonal twentieth-century classical music—the sort of music that breaks all the rules. Paul loved it precisely because it sounded so different. "We were brought up to challenge everything. Conventional wisdom was bound to be wrong, and so if you were unconventional at least you had a small chance of being right," he says repeatedly.

Paul's family bears that out. At the age of nine, his sister Abby cut her hair, called herself "Ab", and joined the local boys' hockey team. She played a ferocious left defence for an entire season before she was picked for the all-stars and her sex was discovered. The story was plastered all over the Canadian press. She was featured in *Time* and *Newsweek*. That was when Paul started calling her Miss Canada. Abby went on to win Commonwealth Gold in the eight-hundred-yard sprint, represented Canada at four Olympic games, and became a famously outspoken member of the International Olympic Committee.

But there are many occasions when imagination is no substitute for experience. Geology is all about weary legs and backpacks weighed down with rock samples. It's about looking at the world you see around you, whether as a record of times past, as an exemplar of the present or as a predictor of the future. To be a geologist is to be rooted in the real world, to go with what you know.

And to some of the people listening to Paul's lectures and seminars, the Snowball was going too far. How could the Earth possibly behave in such an extreme way? He wants oceans that freeze over completely, even in the tropics and at the equator. An ice age that lasted for millions of years. A planet that then plunged from the coldest temperatures it had ever experienced into an intense hothouse within just a few centuries. Carbon dioxide levels hundreds of times higher than have ever been seen in the geological record. Rock weathering rates like nothing on Earth today. How could anyone ever accept a theory that was so far out of the box?

The more Paul pushed, the more vehement many of his critics became. Particularly a certain geologist from New York called Nick Christie-Blick.

NAMIBIA, JUNE 1999

A FLEET of trucks swept off the concrete forecourt of the Safari Hotel in Windhoek and began the long trek north. This was phase two of Paul Hoffman's Snowball mission. After his intensive programme of lectures, seminars and presentations, he now brought a selection of his peers out to Namibia to see the Snowball rocks for themselves. Among them was Nick Christie-Blick, a professor of geology at Columbia University's Lamont Doherty Earth Observatory. Nick was unsympathetic to the Snowball, annoyed by Paul's combative style and dismayed by the implication that the Earth had behaved quite differently in the past. When Paul invited Nick on the field trip, he expected a certain amount of trouble. "I knew Nick would be a pain in the ass," he told me later, "because he always is." What Paul didn't realize was that the field trip was about to turn Nick into the Snowball's Chief Unbeliever.

Geology is an intensely personal science. It's not enough to study a sample of rock that someone has carried home, or even to see photographs that they've taken. In this mind-twisting game of constructing a three-dimensional jigsaw from rocks that have been bent, thrust over one another, eroded away or buried, the context can be everything. Show me just how sharp the contact is between the two rock types on the outcrop, and how far it extends before it disappears from view. Where exactly in the cliff face did these measurements come from? How accurate are your sketches? How detailed are your maps? Geologists usually trust their own field data completely, but are much more reluctant to place reliance on data from places they've never seen for themselves. Plate tectonics pioneer William Menard put it well: "Some earth scientists believe in God," he said, "and some in Country, but all believe that their own field observations are without equal, and they adjust other data to fit them."[10]

That's why field trips make up an important part of how geology is done. You do the field *work* on your own, or with your few closest collaborators, for month after lonely month of mapping, sampling, walking out contacts. When you're back home again, you might write a scientific paper describing what you've seen and adding whatever interpretation you see fit. But the real test comes when you take your colleagues out to look at the outcrops you've been working on, so they can judge the rocks for themselves.

So Paul had brought along experts in Precambrian geology from around the globe. He had arranged everything, even paid for the trip out of his own precious grant money. This, he felt, was the one way to convince his fellow geologists that the fledgling Snowball theory was sound.

Yet Nick Christie-Blick grew more antagonistic every day. Nick, like Paul, is a field geologist. He patrols the world's rock surfaces for a living, measuring and probing and hammering his way through millions of years of history. He lives in the United States, but is thoroughly British. He drinks tea, speaks softly with a clipped Home Counties accent, and has the peculiarly English habit of carefully enunciating some words in a sentence, and then unaccountably rushing and garbling the rest. His short dark hair curls slightly over his forehead and around his temples. He has an engaging, self-deprecating smile. His face is clean-shaven and square, and his build is muscular. He's more football player than long-distance runner. He prides himself on his fitness.

The world of ancient geology is a small one, and Nick and Paul Hoffman have known each other for years. They first met on a Christmas expedition to the Grand Canyon in 1974, when Paul was thirty-three and already a well-established geologist, and Nick was a callow young graduate student of twenty-one, fresh off the boat from England. The North Americans on the trip were kind to their young English colleague. They lent him extra clothes at night when his sleeping bag proved woefully inadequate. They included him in their arguments about geology, bebop and baseball. They called him "Blick".

Paul in particular made a big impression. Even back then, Paul had a reputation as a very talented field geologist who was both a "doer" and a "thinker". Nick was eager to learn from him. But the memory that stayed with Nick most strongly came on the day the group was climbing out of the canyon up the Kaibab trail. Though the climb was steep, Nick wasn't particularly worried.

He'd spent the previous three years rowing for his college in Cambridge. He was strong and fit, an outdoors type well used to holding his own on arduous hikes. He was also used to being first up the mountain. That's why he remembers so keenly the moment when a lean, spare Paul Hoffman overtook him from behind. The trail was as steep as it gets in the Grand Canyon, but Paul was almost running. There was no catching him. As the rest of the group watched in astonishment, he tore up the hill and disappeared. Paul swears he didn't do this for effect. "I just climb fast," he said, shrugging. But effect it certainly had. "He left us for dust," Nick told me. "It was certainly impressive." And he threw back his head and laughed.

At first, Nick hadn't intended to gun for the Snowball theory. He was already involved in too many other arguments. In a way, you could call Nick a professional critic. He is famous for picking away at the threads of theories until he finds a detail that unravels the whole thing. But that kind of criticism takes time and trouble. Nick was forever coming across theories he disagreed with—some were big overarching ideas like the Snowball, others were arcane details of rock behaviour. He wouldn't have the energy to try to disprove them all. "Life is too short," he told me once. And if the Snowball had been fated to disappear back into cosy obscurity, Nick would probably have stayed out of it.

His opposition had grown, though, after Paul turned up at Nick's home institution to give a Snowball lecture. Paul arrived late. He had been caught—ironically enough—in a snowstorm driving down to New York from Boston. One of his tyres had been cut by jagged ice on the road, and he'd had to replace it in the nasty driving sleet at the roadside. The folks at Lamont Doherty had almost given up on Paul when he finally arrived.

Lamont's genteel buildings are set on a lovely old estate just outside New York. It's one of the world's top places for studying the way the Earth works, and it is famous for being fiercely competitive. Scientists there are actively encouraged to comment on and criticize other people's work. Interactions are forthright and robust. Paul knew well that if you take a new idea to Lamont, you'd better be ready for a fight.

So Paul went into Lamont with all guns firing. And as Nick listened to the talk, he became increasingly incensed. Paul used words like "panacea" and "triumph", the sort of words that bring Nick out in a rash. He hates big ideas that purport to explain everything. In his view, they are invariably wrong. And the people who advocate them nearly always end up sweeping inconvenient details under the carpet. That's why details matter so much to Nick. The world, he says, is complicated and the only way to explain it is by laboriously piecing together small parts of each individual puzzle. When you start talking about panaceas, he says, that's the first step towards donning blinkers and losing all sight of what's really out there.

Nick takes his obsession with details out into the field. He has been known to stand up on a cliff top with an ironic grin, throw out his arms, and say: "Hallelujah, come on down, all you believers!" He heard Billy Graham say that once, inviting the converts down to the stage, and it struck him as the perfect metaphor for geologists reverently making their way down to the rock face and the precious clues about the Earth that lay therein. But in reality, Nick is more of a Doubting Thomas. When he goes to see the rocks for himself, he has to put his hands in the wounds. He has to see the processes for himself. Only when every single question, however small, has been fully answered and every doubt satisfied, does he allow himself to believe.

And on Paul's field trip to Namibia, that attitude proved disastrous. Between Nick and Paul there couldn't have been a more dramatic clash of personalities and styles. Nick wanted to find holes in everything. He argued incessantly about every outcrop and every interpretation. Paul, on the other hand, didn't want to know. From the first day and the first outcrop, it became obvious to Nick—and everyone else on the trip—that Paul didn't really want to hear alternative interpretations. Paul had convened the field trip to persuade people, not to hear his theory criticized at every turn. The more Paul refused to listen to Nick's criticisms, the more determined Nick became to find fault.

Nick, in confrontational mode, can be truly infuriating. I first met him in the departure lounge at Las Vegas airport, months after Paul's field trip. By then, Nick was implacably opposed to the Snowball. His first words to me were, "Snowball Earth is dead." He didn't say, "I disagree with some aspects of this theory," or "I think there are certain problems with the interpretation." He said it was dead. The airport was full of geologists on their way to a conference, and many of them were buzzing with the Snowball idea. It was manifestly alive and kicking. But rather than pointing that out, I replied that I'd be interested to hear why he believed that, and mentioned that I too had visited Paul's field site in Namibia. Nick curled his lip. "Oh," he drawled scornfully, "so Paul's taking *tourists* to the field now, is he?" That was the most damning thing he could think of to say. Later he apologized. Though Nick is exasperating in the heat of battle, he can also be humorous and pleasant when he backs off. He said he had just spent the past few days arguing with a long-standing adversary about how exactly to interpret some rock arcana. He had, he explained ruefully, "come out punching".

Paul's response to this kind of behaviour, though, was equally infuriating. Paul has a habit occasionally, if you've said something that he doesn't want to hear, of simply erasing it from the airwaves. He might do this with anything he doesn't want to comment on—an anecdote about someone you know and he doesn't, an opinion he disagrees with, an emotional experience that he can't connect to. When you say one of these things, he doesn't react in any way. He just pauses until you've finished, and then continues with whatever he was talking about before. Rather than ignoring your comment in some pointed way, he behaves in every sense as if you simply didn't say it. Sometimes, talking to Paul, I've caught myself wondering if I really did say something, or just spoke it in my head. It can be unsettling, but it's also relatively infrequent.

But with Nick, Paul began to do this constantly. Many of the other scientists on the trip started to feel uncomfortable. This behaviour seemed just as bad as Nick's continual carping. During a field trip, up on an outcrop, you're *supposed* to discuss things. Dan, the "people person", did his best, trying to engage Nick in just the sort of discussion that Paul was eschewing. But it didn't help. Nick was scornful of Dan's interventions. Dan wasn't a field geologist, and he knew little about rocks as ancient as these. He was no substitute for Paul, and Nick had no qualms about saying so.

On 22 June, just over a week into the field trip, the convoy reached a spectacular outcrop in the northwest of the country. As with many of the best outcrops, Paul had stumbled across this one almost by chance. Two years earlier he had been exploring a dried-out gully, trying to trace the uppermost part of the glacial rocks. The gully was moderately hard going, choked with hefty pale grey boulders. As Paul made his way laboriously up the main channel, a graduate student named Pippa Halverson ducked off down an innocuous and apparently uninteresting side channel. A few minutes later Pippa reappeared. "You might want to come and look at this," he said.[11]

"This" turned out to be an outcrop that took Paul's breath away. To reach it, he and Pippa scrambled down the side channel and then turned the corner that was hiding it from view. The rocky ground rose steeply up ahead, but the surface was so pitted by the action of wind and weather that their boots stuck to it like glue. There was little vegetation, just a few scrubby bushes and arthritic trees, one with a gleaming bark like thickly smeared cream. And rising up on the left was a sheer cliff face that was the embodiment of the Snowball story.

The base of the outcrop, later called "Pip's rock" in honour of its discoverer, was crammed with ice-borne rocks of every shape and colour. These were the so-called dropstones that had been de-livered to the seafloor by ancient icebergs. White and pink and tan and orange, they stood out spectacularly against the dull mud-stone. They were the unmistakable sign of ice. But that wasn't all. Around halfway up the cliff, the scene abruptly changed. Sud-denly the mudstone transformed into a pinkish carbonate that contained no interlopers, no boulders, no signs of ice at all. Below this knife-sharp edge between rock types, the Snowball was in full force. A fleet of icebergs floated on an ancient sea, discharging the rocks they carried into the soft mud on the seafloor. Above this edge, everything had changed. The ice had melted, the sea had boiled with carbon dioxide gas and a milky carbonate rain. In spectacular fashion, these rocks had captured the transition between icehouse and hothouse. This one outcrop encapsulated everything about Paul and Dan's story.

Paul is immensely proud of Pip's rock. When he takes you there, he can't resist building up a sense of drama. He climbs on ahead over the giant boulders of the dried-out gully that leads you to the rock, and as you round the corner to see the cliff face, he is already there, ready to gesture towards the outcrop with a triumphant flourish. And Dan then calls for all present to doff their caps in mock respect for the rocks. Nobody is allowed to use their geological hammers to pry rock samples from the surface. Paul has decreed that this particular outcrop must remain pristine.

Geologists often have their sacred places, the ones that hold the key to their ideas. I've seen researchers take off their shoes when walking on rock surfaces. Though they say it's because they don't want to risk damaging the outcrop, it is a strangely reverent gesture.

Some of the most famous outcrops are sites of pilgrimage, for which geologists compile "lifetime lists". One place on every geologist's list is Siccar Point in southern Scotland, where the father of all geology—an eighteenth-century gentleman farmer named James Hutton—first learned about the Earth's great age. Before Hutton, the prevailing theory of how rocks appeared on the surface of the Earth was called neptunism. This theory held that the Earth was once covered completely with a single vast ocean. Each layer of rocks was formed in the ocean, the most primitive ones first, and the more recent ones last. Eventually the ocean dried up and the rocks have remained the same ever since. This idea fitted beautifully with the biblical notions of creation, Noah's flood, and an Earth that had existed—according to the most literal inter-preters of the Bible—for just a few thousand years.

After Hutton, this biblical interpretation was swept aside in favour of a new, more rational approach. Hutton realized that rocks had been created at different times and in different ways. Some were laid down on the floors of ancient oceans, others cre-ated by volcanic eruptions, and others still by the erosion of mountains, whose pulverized rocks spilled into nearby valleys to create new layers of geological history. And, crucially, the Earth cycled through these processes. What had once been an ocean floor could be thrust upwards to become a mountain, then be eroded into a valley, and eventually be flooded to become an ocean once again. The Earth's surface was continually created and eroded away and re-created again, in a process which Hutton fa-mously said had "no vestige of a beginning—no prospect of an end". Through his insights, Hutton had laid the foundations of the notion of deep, unfathomable geological time. His dear friend, the professional mathematician and amateur geologist John Playfair, described the vision of geological eternity thus: "The mind seemed to grow giddy by looking so far into the abyss of time."[12] Of the perception that the Earth had existed for dizzy-ing eons, the late Stephen Jay Gould said that "all geologists know in their bones that nothing else from our profession has ever mattered so much."[13]

But even the rational Hutton obtained his inspiration from re-ligious convictions. In his farming days, Hutton noted that soil was created when old rocks were eroded away, with the debris carried ultimately off to sea. If this were the only process allowed, all of Earth's land would ultimately erode away and there would be nowhere left for mankind to live. Since Hutton believed that a kind and loving God had created the world expressly for the benefit of its human occupants, he reasoned that there had to be another process that rebuilt the Earth's surface and kept it comfortably habitable. That's how he developed the idea that seafloors could become mountains, and that volcanoes could create new land to replace the land that had washed into the oceans.[14]

Hutton's arguments about God's motivations would hold no weight in modern geology, but they show that science is muddier than it seems, and that scientists' ideas and inspirations can come from unexpected sources. What distinguishes science from pseudoscience is not whether your theory originated with some particular conviction about how the world works, or whether you feel an emotional attachment to it. What matters is the evidence you find to support it, and whether you are ultimately prepared to accept that it could be wrong. Perhaps it's appropriate, then, that students of geology flock to the site of Hutton's original inspiration with a most irrational reverence. They go for the sheer pleasure of witnessing first-hand the rocks that inspired it all.

It must have been obvious to everyone on the field trip that Pip's rock was sacred to Paul. And Nick was exasperated. This outcrop wasn't informative so much as photogenic. That, Nick felt, was why Paul was revering it. It was just showmanship. Nick marched up to the rock face and looked for some fault to find.

He found something almost immediately, in the stones that Paul claimed had been dropped on to the ancient seafloor from icebergs floating overhead. You can tell when a stone has come from an iceberg because it deforms the soft mud that it lands in. Instead of lying flat, the layers of mud immediately beneath the dropstone are squashed downwards.

But this should only happen to the sediment *below* the stone. If you peer at the cliff face and see lines of sediment deformed above as well as below the boulder, that's a warning sign that the stone may never have been dropped at all. Instead, it probably rolled down a subterranean slope. And then when the soft sediments were squeezed around the hard boulder they bent around it, top and bottom. Crudely put, distortion below an embedded boulder implies a dropstone, while distortion both above and below implies something else. That's what Nick was looking for at Pip's rock. He moved along the rock face peering at the boulders until he found one to be suspicious of. Look! he shouted in triumph to anyone who would listen. There's deformation above this one as well as below. That's compaction! Look. That shows that some of the boulders didn't come from ice.

Some of the boulders didn't come from ice. But even Nick accepted that many of them did. In other words, this was a pointless criticism. Pip's rock had been created in the presence of ice. Even if a few of the boulders it contained were not dropstones, the rest of them clearly were. Nick wasn't claiming that there had been no ice when the rocks were formed. He was simply pointing out a slightly different interpretation for a few patches of the cliff.

Didn't Nick realize that if he criticized so pointlessly, it would drive Paul mad? "Yes," Nick told me later. "But that's too bad. The point of the field trip was to have people come and take a look. He knew what he was getting."

By the end of the trip, the battle lines were drawn. On the long flight home from Johannesburg to New York, Nick wrote Paul an eight-page, single-spaced e-mail detailing all the arguments and criticisms that he felt hadn't had a proper hearing on the outcrops. In the opening, he was the soul of politeness:

Once again many thanks to you and all those involved in the excellent excursion. . . . I very much appreciate the opportunity to see these fascinating strata first hand, and also the effort you made to obtain financial support for the trip.

But it didn't take long before Nick was launching into detailed criticisms that seemed almost calculated to enrage. He sent the message to Dan, Paul and everyone else who'd been on the trip, and quite a few people who hadn't been—something that Paul later claimed had been done purely to blacken the Snowball's name among people who hadn't seen the rocks for themselves. To Nick's chagrin, although a few of the recipients wrote back to say thanks for the insights, nobody took up his invitation to engage in further discussion. Paul outwardly ignored the e-mail, and inwardly seethed.

The following spring, Nick taught a graduate class at Lamont about the Snowball, and followed up with an e-mail to all the students outlining his criticisms, and warning them that Paul was "a great salesman". Inevitably enough, the e-mail found its way to Paul, who was seriously stung. What he objected to most of all, he said in a heated message to the course's organizer, was the way Nick seemed to want to pass him off as an ideas person who paid no attention to detail. This, he declared, was patently untrue:

Anyone who doubts I have the ability and the will to walk the extra mile at the end of a long day to get the facts right should try me out in the Boston marathon some year.

Nick had no intention of running against Paul in a marathon. But he did decide to take the battle to enemy territory. In September 2000, Nick went to the Massachusetts Institute of Technology, just down the road from Harvard. He had chosen a deliberately provocative title for his seminar: "The Snowball Earth Hypothesis: A Neoproterozoic Snow Job?" The convener insisted on the question mark. Nick hadn't wanted it in. During the lecture, Nick's main criticisms centred around how Paul and Dan were presenting the Snowball idea. It was a cottage industry, he said. A bandwagon. Paul and Dan were in the audience, and both were furious.

Nick and his fellow contrarians are as important for scientific progress as the people whose new ideas they challenge. This process of putting up and knocking down can be one of the best ways to find out whether a theory really holds, whether parts of it need to be massaged, or whether the whole idea should be dropped.

Still, the heat of the Snowball interchanges had its inevitable effect. The aftermath of the "Snow Job" seminar was exactly what Nick must have predicted. There was no more talking with Paul. Nick had defined himself as an enemy of the theory, and there was no going back. Now it was time for real scientific challenges to take over from the rhetoric.

SEVEN

DOWN UNDER

Paul Hoffman had woven together the Snowball story's different strands. His theory was new, but it also rested firmly on observations and ideas from the past, especially those of Brian Harland and Joe Kirschvink. With his carbonates and isotopes, however, he and Dan had provided the first evidence that the Snowball might be right.

Now the theory was about to face its first serious challenges. They came, like so much in this story, from the rocks of South Australia. After Joe had performed his magnetic magic on Flinders rocks in the 1980s, many other geologists had been back to poke at the outcrops, and prize out more of their data. By the time Paul and Dan published their Snowball theory, there was already a stockpile of Australian data just waiting to be mined. And from that data, two tests of the Snowball quickly emerged.

SNOWBALL EARTH

Since Nick Christie-Blick had become the main focus of much of the Snowball resistance, it's appropriate that one of those tests came from his own research group. Ironically, though, the result didn't conflict with Paul's theory at all; quite the contrary. Working in the Flinders, a student of Nick's had already uncovered some evidence that turned out to be greatly in Paul's favour. She had addressed the issue of timing. For Paul and Dan's explanation to work, the ice had to last many hundreds of thousands or even millions of years, long enough for carbon dioxide to build up in the atmosphere and set the conditions for the global layer of carbonates to form. So were they right? Just how long did the Snowball last?

BENNETT SPRING, SOUTH AUSTRALIA, 1995

LINDA'S ARMS were already starting to ache. But she couldn't let go. If the drill doesn't go in straight, the core is ruined and you have to start again. Water, milky with rock dust, was spurting out of the hole and spraying her jeans from the thigh down; they were already clammy and soon would be sodden. She shifted angle awkwardly, trying to rest her right arm on her knee. Even through ear plugs, the noise was deafening.

To her right and left ran a dried-out stream cut, its steep walls casting her into shadow. Though Australia was on the fringes of winter, the temperature had crept into the seventies. The sky was scattered with bright cirrus clouds. Thirty yards to the right, splashes of brilliant green rushes marked a sluggish spring—the only water in this parched country for miles around. Linda had surprised a grey kangaroo there that morning; it had made its characteristic "shhht" sound of alarm and thudded hastily away. Here and there along the stream were stately red river gums, leaves still green, grey bark peeling to reveal the silver trunk beneath. One good reason not to camp in the stream cut—you never know when a gum tree will decide to shed a branch.

Linda's camp was a hundred yards back, up on the flat, surrounded by dry grass and nondescript scrubby bushes, their leaves a dusty grey-green. This was the centre of the Flinders Ranges, a few hundred miles from the rhythmites of Pichi Richi Pass. The ground was a rich golden brown, and the limestone hills around the plain were low, their slopes gentle.

Her vibrant red Suzuki truck, with its soft top, clashed nastily with this muted countryside. But the rest of the camp blended. A pale wooden table with folding legs, a hefty plastic water carrier that held 25 litres, and a tidy pile of twigs and small branches collected in advance for the night's fire. (Always collect wood from gum trees—the few cypresses and pines leave a sticky black goo on the bottom of your pans.) A half-dome tent, Outback colours, green and grey, carefully upwind of the fire.

The bulky HF radio Linda had left in her truck. That had been a waste of limited space. It was a loan from the mines department and was supposed to be working, but all she could hear was static. So much for communicating with the outside world. The only human contact she'd had for a week had been two days earlier when, bumping her way in the Suzuki past the main buildings of the sheep station, she'd stopped and asked permission to put holes in the rocks. The men at the station had stared at her, baffled. Why should she want to? Why should they care?

Linda Sohl, a girl from the Bronx, was finally beginning to feel at home in the Outback. This was her third field season, she was nearly halfway through her Ph.D., and things were looking good. She was twenty-eight years old, plump, pretty, with neatly manicured fingernails, tiny silver hoops in her ears, soft brown hair cut and flicked into symmetrical waves that framed her face. She had enormous deep brown eyes. In different circumstances, without the safety goggles and the field gear and the rock drill, you'd probably take her for a sensible older sister. She'd seem cautious, level-headed, often reserved. But she had an air of resolve about her and—on occasion—the unmistakable look of a dreamer.

Dreams of adventure had taken her two years earlier from her well-paid but dull job at a New York publishing house and propelled her back to college. Her parents were horrified. A Ph.D.? In geology? Where's the security in that? Still, in spring 1993, Linda had arrived in the office of Nick Christie-Blick, whom she'd met at a seminar.

Nick had invited Linda to visit his lab, in the leafy New York suburbs, and Linda was feeling hopeful. She knew how hard it was to get into a Ph.D. programme at such a prestigious place as Lamont Doherty, especially as she hadn't taken the conventional route straight from college. Still, Nick had expressed interest, and that was a good sign. Linda didn't know quite what to expect—an interview, perhaps a tour of the lab. She was determined to make the best possible impression.

No need. Nick had already decided to take her on. When she walked through the door at nine-thirty in the morning, he had spread out various geological maps of South Australia and he immediately began suggesting different field areas. "Do you have a passport?" he asked her. "We need to get you a visa." Several weeks later they were driving together through the Australian Outback.

She'd never camped before, not even with the Girl Scouts. She'd never even seen a sheep before, not close up at least.

A few days after they set out, Linda and Nick were in the car park of a small motel, transferring gear from one vehicle to another. Back and forth they went, arms full, past a pickup truck loaded with rusty old farm equipment, bales of wire, and a sheep—lying on its side. Linda was fascinated. She stopped and stared. It didn't blink, didn't seem to be breathing. It was absolutely still. Its eyes were a dull light brown colour. She thought they looked like the blank, dead eyes of a stuffed teddy bear. She put her head closer. Suddenly, without warning, the sheep blinked and turned its head. Linda yelped and leapt backwards. "Nick! The sheep's alive!" Damn, damn, damn. Not the way to impress your brand-new adviser, before you've even made it into your first semester.

Nick was unfazed. He bought her a tent, a lantern, a set of pans. He took her to various possible research sites in the Flinders Ranges. He helped her to choose which rocks she wanted to work on for her Ph.D., showed her how to hook up the radio, and left her there.

At first the silence was the worst. Then the unexplained bumps in the night. Kangaroos would hop silently by until they hit a gravelly streambed and suddenly wake her with their crash. Reclusive emus would come out at night to skitter on the stream pebbles, left, right, left, right, like people running. Once there really were people, whooping their way along a remote bush track, shining bright searchlights and firing guns. The air was full of bullets and agonized screams. Linda stayed in her tent. Later she discovered it was part of an authorized cull of the feral alien animals, the cats and foxes that bedevil the Outback.

But for the most part the Outback turned out to be benign and beautiful. For annoyance value, there were just the idiotic galahs, brightly coloured parrots that would periodically launch themselves en masse from a tree, shrieking wildly and dive-bombing the site. And the bugling cockatoos, and the oversized Australian magpie, actually a relative of the crow, but with an en-chanting, lyrical warble that echoed hauntingly among the gum trees, and almost made up for the rest. The birds were useful in a way. No need for an alarm clock when you have an ear-shattering dawn chorus every morning, 6:45 sharp.

Linda had never intended to test the Snowball idea. She was supposed to be studying some carbonate rocks that had formed long before the ice. But, having worked among the ice rocks for the past two years, she couldn't help being intrigued by them. She'd heard about Joe Kirschvink's magnetic work and the more detailed studies by George Williams. And she decided to see whether she too could detect the faint magnetic traces in the rocks.

A few weeks into her 1995 trip, Nick came out to help. He hadn't encouraged Linda to work on the ancient ice, but he was curious to know how her results would turn out. Nick, the details man, was always pestering and probing. In the field he could drive you mad. He doesn't—ever—let something lie. As he and Linda drilled and washed and carried samples back and forth, he asked her incessantly about the precision of her magnetic mea-surement. To Linda, it simply wasn't the most important issue. Details often matter, but that particular detail would come out in the wash. But Nick was consumed by it. A few weeks after he left to return to his own field site, Linda went into the small town of Hooker to collect her mail, sent to her care of "general delivery". Among her letters was one from Nick, via his field camp several hundred miles to the north. The page was covered with careful diagrams and a detailed analysis of the factors affecting the precision of Linda's measurements. He just couldn't let it go.

At the end of the season, Linda headed back to New York to analyse her samples. The work was exhausting and painstaking. The magnetics machine was already booked up for the days and evenings, so she worked on it through the nights for six weeks. But in the end she hit pay dirt. Though all the rocks she collected had looked more or less the same, pale and reddish from the magnetic minerals they contained, some of them had a field pointing northwest, and in others the field was southeast. To Linda, that meant only one thing. These ice rocks contained a record of ancient magnetic reversals.

A magnetic reversal occurs when the Earth's magnetic field spontaneously swaps directions, so that magnetic north becomes magnetic south. Our planet's magnetic field is generated by molten iron sloshing around in the Earth's core, and the flipping must somehow be related to changes in this deep, hot liquid. But the details remain a mystery. What we do know is that this bizarre event takes place roughly once every few hundred thousand years. If you could slow down your frenetic human lifespan until it matched the stately passage of geological time, you could study a compass and watch these flips in action. Flip, and the needle would swing halfway around the dial to point south; flip, and it would point north again; flip, back to south. The magnetic minerals trapped in Linda's rocks had behaved exactly like that geological compass needle. Each ancient flip had been frozen in.[1]

These flips are useful to those who study the ancient Earth, for two reasons. First, if they appear in a sequence of rocks, the magnetic field there has to be original. Neither subsequent heating of the rocks nor the influx of new magnetic material can produce this alternating pattern of fields. By finding reversals in the Flinders ice rocks, Linda confirmed Joe's discovery that ice had been present near the equator.

More important, Linda had also found evidence that the ice was extremely long-lived. Her rocks contained perhaps as many as seven flips. If magnetic reversals in the Precambrian happened roughly as often as they do today, Linda's ice rocks had to span at least a few hundred thousand years, and probably several million. Here was the first tangible evidence that the Snowball glaciations really were the longest ever known. It seemed as though they had lasted quite long enough to build up huge amounts of carbon dioxide in the atmosphere, and trigger the events that Paul and Dan envisaged.

Paul was thrilled when he heard about Linda's reversals. The Snowball theory had survived its first important challenge—the test of time. Nick was perfectly aware of the irony that his own student's work had ended up supporting the Snowball, but he was also proud of Linda. She had made an intriguing observation, providing a potentially important clue about the conditions that prevailed during the glaciations.

Still, just because the ice had stuck around at the equator for an inordinately long time, that didn't mean the rest of Paul and Dan's theory was correct. Nick and Linda were both still determined to show that the overall picture was nothing like Paul and Dan's dramatic total freeze-over.

The next serious test of the Snowball, however, wasn't from Nick's camp, though it did come with another dose of irony. Once again the challenger had worked in the Flinders Ranges, and provided key evidence that was later incorporated in Paul's theory. But this wasn't Linda, or anyone who worked with her. The new challenger was George Williams, the Australian who had instigated the research that caught Joe Kirschvink's attention back in the 1980s. George had worked in the Flinders Ranges for decades. He's the person who studied the tidal rhythmites there, and whose work laid the foundation for all the magnetic work that followed. He had done plenty of further magnetic work there, too, and he was utterly convinced that there had been ice at the equator. But he was also convinced that this had nothing to do with a global Snowball. George had an alternative explanation: the Earth, he said, had tipped over on to its side. Sound crazy? Well, he had evidence that seemed to prove it, and that was soon giving Paul Hoffman sleepless nights.

PORT AUGUSTA is a small, grim coastal town a few hundred miles north of Adelaide, just west of the Flinders Ranges. Its street signs proclaim it to be the "Gateway to the Outback". At Port Augusta the rough stuff begins. Though the roads to the south are smooth and civilized, many of the northern ones are little more than dirt tracks.

From there, the driving instructions to Mount Gunson Mine are simple enough. One paved road goes north, the Stuart Highway. Take it. After a hundred miles or so, turn right. Yes, this will be the only right turn. You can't miss it. No, really, you can't possibly miss it. There's nothing else there.

Most of the trade between Port Augusta and the Australian interior is plied along the narrow Stuart Highway by terrifying, thundering "road trains". These are linked caravans of two or maybe three huge trucks pulled by just one engine, whose driver has probably been up all night, fuelled only by greasy steak sandwiches "with the lot" (onions, cheese, tomato, bacon, fried eggs, you name it, all oozing with ketchup and evil yellow mustard) from the occasional wooden roadhouses. The trucks rock your car with shock waves as they rip past you. They are death to kangaroos, and grey-furred flesh periodically smears the road, picked at by huge, black, wedge-tailed eagles. Even a full-sized kangaroo won't make much of a dent in a road train. But if you're in a regular car, you'd better not drive at dusk or dawn when the 'roos are on the move. Evolution hasn't had time yet to equip them with road sense. They'll lurk in the scrub by the roadside and then leap into your path without warning. Hit one, and it could well be the death of you, too.

There's little else to distract you during the drive, just the flat, empty, featureless scrub of the Australian desert. Eventually a slightly battered sign points to a dirt road on the right. These days not many vehicles take this turning. In the 1980s, Mount Gunson was a flourishing copper mine, but now most of the operations have closed down. A few workers' huts remain, with their peeling girly posters. Also some yellow warning signs—SULPHURIC ACID. CORROSIVE!—next to rusting pipes almost the same colour as the dusty red soil. But the rest of the site is scarcely different from the surrounding desert, its landscape scattered with drab grey saltbushes and Flinders Ranges wattles, fast-growing, fast-dying, dried-out skeletons rubbing up against the vivid bottle-green of the living. When the miners finally pull out completely, the only sign of the once-frenetic mining will be the great open pits carved deep into the ground. Now that the mining has ceased, water is no longer pumped out of these pits, and soon they will become lakes. Then, there will be no more chances to see the geological signs that caused Paul Hoffman such anxiety.

The Northeast Pit, cut into the Cattlegrid ore body, bears the features that seemed to challenge Paul's Snowball. This great hole in the ground, a thousand feet long and almost as wide, sinks down to a flat floor. The sides are steep and layered, dark quartzite near the floor of the pit, paler sandstone above, and then, in the topmost layer, the ubiquitous iron-rich red of the soil. You crunch along the floor of the pit on delicate yellow nodules crusted with white crystals of gypsum. Patches of soil are slotted with short parallel lines, where passing kangaroos have dug in their heels. The sun is invariably dazzling, glaring off the sandstone. Squinting even through sunglasses, you have to be close to the walls before you notice anything different about them. But as soon as you pick out one of the strange structures in the pit's steep walls, you start to see them everywhere.

They are wedges of sandstone, six feet tall and triangular, like a row of massive shark's teeth. Or perhaps witch's teeth, since they are stained green by the flow of copper-rich water. The quartzite they sit in has been smashed and broken like a jumbled pile of bricks, but the wedges themselves, and the rock layers above them, are made of smooth, flowing sandstone. They line the walls of the pit. As you walk along you see first one, then another, then a whole row of them, strung as on a necklace. Occasionally you see the outline of a wedge or two above the main row. To an untrained eye they look bizarre. But to geologists they're classic. They're textbook. Anyone can tell you what they mean.

Sand wedges are the clear signs of a climate pattern called freeze-thaw. Here's how they form. First, freeze. The temperature drops quickly, and in response the ground shrinks and cracks into regular polygonal shapes, like the mud cracks that form in the bed of a dry lake. Into the cracks blow sand and dust. Next, thaw, and the cracks are kept open by the presence of the sand. Next, freeze again, and the cracks open wider, with more sand tumbling in. Eventually the sand that has wedged into the cracks solidifies and turns into sandstone. In Mount Gunson's Cattlegrid pit, the broken-up quartzite is the ground that was repeatedly cracked by freezing and thawing, and the shark's teeth are slices through the solidified sand wedged into the cracks.

So Mount Gunson must have suffered from repeated episodes of freezing and thawing. Why should that be a problem? Mount Gunson was an island then, surrounded by a frigid ocean. Over to the east are the ice deposits of the Flinders Ranges, where fleets of icebergs dropped their load of stones and boulders into a shallow sea. Freezing the ground thereabouts should have been easy.

But to get sand wedges doesn't just require cold temperatures. It also takes warm ones, followed by sudden, repeated temperature drops. Sand wedges need cycles of freeze and thaw—in other words, *seasons*. The problem is that Mount Gunson was close to the equator when the ice was present. And at the equator, seasons simply don't happen.

We have seasons because our planet is tilted. If the Earth remained bolt upright in its progress around the sun, there would be no such thing as summer and winter. All year round, every place on Earth would experience the climate that its local position deserved. Close to the equator, where the sun is fierce overhead, the climate would be hot. Close to the poles, where the same amount of sunlight spreads over a larger area, the climate would be cold. January or June, there would be no difference.

But the Earth's tilt makes life more interesting. Superimposed on the overall pattern of climate—hot equator, cold poles—is a seasonal shift. In January the southern hemisphere is thrust out towards the sun. Australians and South Americans head to the beaches. Antarctica basks in the midnight sun, and temperatures there can stay above freezing for days. By June, halfway again around the Earth's annual orbit, the northern part of the Earth receives more than its share of sunlight. Now Antarctica is shrouded in permanent darkness, and northerners take their turn in the sun.

The equator is the only place on Earth to escape this annual cycle of hot and cold. No matter which hemisphere is grabbing its extra share of sunlight, the equator feels it, too. Equatorial regions muscle in on everyone else's summers.

So how to explain the Mount Gunson sand wedges? Cast-iron magnetic evidence says that Mount Gunson was at the equator when the ice came. Nobody doubts this. But the wedges seem to show seasonal changes. What gives?

George Williams thinks he knows. Today the Earth's axis isn't tilted by much, just twenty-two degrees, a sixteenth of a full circle. But, says George, what if it used to be much more tilted? Perhaps, in "Snowball" times, the Earth had tipped over on its side, by something closer to a full quarter-circle.

If so, everything we now know about the climate would be turned upside down. Poles and equator would swap characteristics with the sun blazing directly overhead at the poles, and spreading feebly out at the equator. And this, George felt, would nicely explain two of the main conundrums of the ice era without requiring a frozen Earth. The Arctic and Antarctic regions would be balmy, and the equatorial regions frozen, which would explain the Australian evidence for ice at the equator. What's more, the sun would now be wreaking seasonal changes on to the frozen equator. That would explain the sand wedges. If George is right, the Snowball simply didn't happen.

George had been talking for decades about the Big Tilt.[2] But until Paul came along, nobody was particularly listening. When George heard about Paul's Snowball, however, he went immediately on the offensive. In a magazine called *The Australian Geologist*, he published a ten-point criticism outlining exactly why he thought Paul was wrong and the Big Tilt was right. (George called this criticism "Has Snowball Earth a snowball's chance?"[3] Paul's immediate response was wryly titled "Tilting at Snowballs".[4])

The Big Tilt itself, after all, faced plenty of problems. For one thing, it couldn't explain nearly as much as the Snowball purported to. The Tilt accounted only for the equatorial ice and the strange sand wedges. It said nothing about the cap carbonates, the isotopes or the ironstones.

More seriously, as Paul quickly pointed out, there was no easy mechanism for righting the Earth. Tilting the planet in the first place would have been quite straightforward if it had happened early enough. When the solar system was born, there were plenty of planet-sized chunks of rock flying around the place, crashing into each other like giant pinballs. Most scientists believe that an almost-planet the size of Mars smashed into the young Earth, creating our moon from the debris. A collision like this could easily have knocked the Earth over on to its side. But in the relatively sedate time of the past few hundred million years, what could have righted the Earth to the more gentle tilt that it bears today? By the end of the Precambrian, any builder's rubble from the early solar system was long gone, and there's no other easy way to move the Earth back upright.[5]

Of course, saying we don't know how the Earth could have

righted itself doesn't prove George's theory to be wrong. Until recently nobody knew what exactly was causing the continents to drift, though they were still moving for all that. But the Big Tilt had a bigger problem, one that Paul seized on gleefully. The evidence for glaciation comes in one particular time slice, right at the end of the Precambrian. So George couldn't rely on the creation of the moon to knock the Earth over. Instead he had to propose that something else had suddenly tilted the Earth around 700 million years ago, billions of years after the formation of the moon. And that something else again had caused the Earth to right itself abruptly again a couple of hundred million years later. That's an even bigger stretch of the mechanism problem.

But there still remained the vexing issue of the sand wedges. How did a place that was within perhaps ten degrees of latitude of the equator experience seasonal changes? Paul did his best. He came up with a possible explanation involving glaciers that alternately surged forwards and drew back, sometimes insulating the ground they covered from the bitterly cold air and enabling it to thaw a little, sometimes exposing it for another freeze cycle. Not many people were persuaded by this, though. It sounded rather too much like special pleading.

Salvation came from a different source, and showed that even Paul had been too conditioned by how the Earth works today. The key turned out to be how different the Earth's seasons would have seemed on a planet that was blanketed in ice.

By 2001, veteran climate researcher Jim Walker, from the University of Michigan, had become intrigued by Paul's Snowball theory, and began tinkering around with a simple climate model to try to figure out what the weather would have been like. He picked the most extreme of Paul's conditions—a globally frozen ocean—and set the model on its way. Day to day in Jim's model, the weather was rather boring. Nothing much changed. There were no travelling storms and no temperamental weather patterns. At every point on the surface of Jim's model planet, one day was pretty much like the next. The Snowball must have been a little like Mars, he says. Apart from the occasional dust storm, the whole Red Planet just settles down into a placid, predictable weather pattern. "The wind there always blows from the same direction at four o'clock in the afternoon."

But to Jim's astonishment, the seasons on his model Snowball were an entirely different matter. They were exaggerated, larger-than-life versions of the seasons we are familiar with. At any one point on the frozen surface, there was a *huge* temperature difference between winter and summer, much bigger than we see today.

Why? Well, our modern wet and windy Earth has an ingenious built-in mechanism to guard against extremes. If the climate everywhere simply depended on the direct overhead sunlight, the tropics would be much warmer than they are today, the higher latitudes would be much colder, and the seasons would be very much more marked. Instead, the Earth's oceans damp down the sun's ardour.

All through the summer, the oceans soak up sunlight. Unlike land, the oceans are pretty much transparent, so the sun's rays can penetrate deep into the interior. Also, the ocean's currents keep water moving. Warm surface water is replaced by cold water rising up from the deep to take its place in the sun. The oceans work like a vast storage heater: they absorb heat throughout the summer, and then slowly release it during the winter. That's why seasonal changes are so much more extreme in the middle of continents than they are at the margins. Places like Nebraska or central Siberia are too far from the ocean to benefit from its summer cooling and winter warming.

But on the Snowball, all that would have changed. According to Jim's model, the whole place would have been like Nebraska or Siberia. With ice covering the oceans, there would have been no more of this gentle amelioration of the seasons. And here's the key point: that argument would apply right down to the tropics. Even places just a short distance from the equator would have had exaggerated seasons. The Flinders Ranges of South Australia would have experienced a temperature difference between winter and summer of perhaps 30 degrees C. And though the annual average temperature would have been bitterly cold, summer temperatures could even have crept up above freezing for a few brief months of the year. Bingo. There's your freeze and thaw. There's your explanation for the sand wedges.[6]

George Williams still believes in his Big Tilt, but most other researchers have begun to edge away from it. The Snowball, it seemed, had survived yet another test. But Linda and Nick were still working on finding an alternative idea, one that was less wacky that George's and less extreme than Paul's. Nick's final e-mail to Paul, back when they were still—just about—speaking, had contained this jaunty assurance: "In parallel with an effort to develop good tests, I also accept your challenge to seek a better hypothesis." And that's exactly what Nick did. He and Linda teamed up with an Australian researcher and longtime collaborator of Nick's, Martin Kennedy. And by the end of 2000, it was beginning to look as if they were on to something.

EIGHT

SNOWBRAWLS

"How can *anyone* look at these deposits and *still* be talking about a Snowball?" Martin Kennedy thumped his fist against the dashboard in frustration. He was in a truck with several other geologists, heading north towards the long, thin ribbon of Death Valley in California, and as the ground fell further and further below sea level, Martin's blood pressure was rising.

Martin is tall, slender, half Australian and half American. He is thirty-eight. He's intense and ornery, but this is coupled with bursts of humour that make him unexpectedly good company.[1] He can launch into a sudden fury, and then pull himself out of it just as quickly. (During a brief period working at Exxon, he underwent the regulation psychological tests, which determined that he was a "red" person, categorized as highly aggressive. He was, he says, disappointed. "I was kind of hoping to be blue-green.") His hair is short, brown and slightly curly. The corners of his eyes are crinkled with laughter lines. Martin has boyish features, a button nose and thin lips that can make him look engaging or petulant, depending on circumstances. He has a pathological aversion to authority.

Martin hadn't meant to pursue a career in geology. Initially he had intended to run a farm in Australia, where "you can live by your own hand". He would probably have done so if the purchase hadn't fallen through. Though he's now based in the relative civilization of the University of California's Riverside campus in Los Angeles, he's happiest in the Australian Outback, where he's done fieldwork for many years. He trusts himself, resents interference, and instinctively challenges received wisdom.

Martin had become increasingly outraged by the claims that Paul and Dan were making for their Snowball theory, and was now a key player in Nick Christie-Blick's stated quest to find an alternative.

In November 2000, he had joined a small field trip that was visiting some Snowball rocks in Death Valley. The trucks headed north through the valley, hugging the east side of the central salt-pan. At first the sand was hazy with low, olive-green saltbushes, but the vegetation gradually disappeared, and by the time the trucks reached the sluggish saline pools of Bad Water, the lowest point below sea level in the western hemisphere, there was just bare sand streaked with salt.

Lining the sides of the valley were mountain peaks, sometimes jagged, sometimes rounded. From a geological perspective, all the mountains are recent upstarts. Around 13 million years ago, a mere snip compared to the Snowball timescales, this part of the Southwest was fairly flat, dipping off the back slope of the Sierra Nevada. Then the land began to stretch. As the crust thinned, cracks opened in its surface. Some parts fell downwards to make long, thin valleys, while others were thrust upwards into mountains. The whole area today is riddled with these stretch marks. And thanks to the overhead thinning, volcanoes sprang up from the deep and spilled their magma on to the Earth's surface. Many of the mountains left from that turbulent time are piled with lava, the rocks painted in a desert palette of burnt sienna and ochre, chocolate, Venetian red and tan.

The Snowball rocks that Martin had come to see hailed from a time long before the Earth's erratic bucking threw up these mountains and opened these valleys. They come not from 13 million years ago, but 600 million years ago and more, eons before dinosaurs roamed North America, in fact before anything roamed anywhere. When the Snowball gripped the Earth, the only living things hereabouts, and anywhere else in the world, were those tiny, single-celled sacs of chemicals bound together by extruded slime.

Though Death Valley is full of Native American trails, it wasn't seen by white men and women until the mid-nineteenth century, when the first pioneers passed through on their way to hunt for California gold and glory. Its moniker is unfair. The valley wasn't so dangerous, even in the early days, if you could find the waterholes. Few people died here, though many had miserable crossings. The landscape is bright with sand and salt and sun. And in spite of the generations of geologists who have picked over the valley in the past century, it is still a place of geological mysteries. Like the migrating boulders of Racetrack Playa, a dried-out lake bed to the north and west of Bad Water. The floor of this valley is strewn with rocks of every size: pebbles, stones and vast boulders weighing seven hundred pounds or more. And, inexplicably, these rocks move. Try to catch them at it and they will just lie there, solid, immovable and innocent. But between visits from the geologists who carefully plot their positions, these restless boulders somehow skid along the valley floor, twisting, turning, zigzagging, and leaving grooved trails behind them. There have been many attempts to explain this bizarre Death Valley phenomenon. Some people think the culprit is an intense gust of wind occasionally funnelling through the valley; or a sudden rainfall that coats the mud with a slippery sheen; or thin sheets of ice that lift the boulders and make it easy for them to slide. Nobody really knows.

The trucks turned west now, where a wide sweep of open land sloped toward the sand dunes in the centre of the valley. Except for the small pools at Bad Water, the valley floor was bone-dry. But up over the western mountains, occasional snow clouds hovered like pale will-o'-the-wisps, trailing off at their base where the snow evaporated before it ever hit the mountain surface. As Martin's truck wound gently up the road through Emigrant's Canyon to the mountain pass, a few fat flakes of snow smacked the windscreen.

Over the pass, walls of rock stretched upwards at the roadside, speckled with coloured rocks and boulders. This was the Kingston Peak diamictite, a direct remnant of the Snowball. The mismatched mélange of rocks had tumbled into the ice-covered seas of Snowball time and left behind a deposit more than two miles thick. Glancing at the rocks, Martin began to get annoyed. "How could you get all this in the final days of the Snowball?" he demanded, gesturing out the window. "If you believe Paul's Snowball, all this was deposited in a thousand years or so. You just can't get sedimentation rates like that!"

This was part of an ongoing argument. In recent months, many people had been wrestling with the question of when exactly the ice rocks had formed. Was it only at the end of the Snowball, or were they continuously created throughout the previous several million years that it gripped the Earth? It mattered because of thick deposits such as these. Everyone agreed that they were created when glaciers gathered up rocks and dragged them to the sea, or when icebergs melted on the water and dropped their load of debris on to the mud below. But in the first incarnation of Paul and Dan's Snowball idea, this couldn't have happened during the long, cold millennia of the Snowball. When the oceans are frozen over, icebergs can't break off and move around, because there's nowhere to move to. And with frozen oceans, it's extremely hard to get ice on the land. To make a land glacier you need snow, and to make snow you need some patches of open water to deliver the moisture. At first Paul and Dan believed that the Snowball would have been cold, dry and dead, with no snow, no glaciers and no icebergs until, perhaps, the very final days when the ice began to melt back.

Then how to account for two vertical miles of rock? Martin was right—you couldn't get such thick deposits from such a short time. But Paul's Snowball idea had evolved since he first started working on it. Now he was allowing for some glaciers to form and move even while the oceans were still frozen. He had realized that the wind could erode ice from the sea surface, transport it to land, and deposit it there. Though the process of creating glaciers would be painstakingly slow, during the Snowball there was no shortage of time. When you have millions of years to play with, it's not so hard to create a river of ice, inch by inch.

Martin knew about this argument, but he wasn't impressed. He switched to another complaint. "What about the cap carbonates?" he demanded. "To make them in the short time that Paul and Dan want, you'd have to have weathering rates a thousand times faster than today's. It's impossible!"

Then, unexpectedly, he grinned. "I feel like I'm getting riled here, and I shouldn't be," he said. "Look, to be honest, I hope the Snowball's right. It's a beautiful idea. But I just don't like the way they're ramming it down our throats. I feel . . ." He hesitated, searching for the right word. "I feel violated."

Martin had first met Nick Christie-Blick, the Chief Unbeliever, back in 1993 on a field trip in central Australia, which had been organized by several senior Australian geologists. Martin was in a foul mood. Even though the group had travelled to his field site, the place he had spent his Ph.D. mapping, Martin had been the last to hear about it.

Geologists can be very protective of their field sites. They spend months there, often alone or with only a few others for company. They leave their bootprints on the soil, and the marks of their hammers on the outcrops. Day after day they climb cliffs and hike through gullies, walking out the contacts between rock types. They learn how every rock and stone is related. In their heads and their field notebooks they gradually assemble the complex, four-dimensional jigsaw that tells them the area's ancient history. And they don't just become experts in the rocks; they often also develop a physical, almost proprietary connection to the landscape. If you plan to visit someone else's field site, the first thing you'd better do is call them.

Martin was no exception. He had grown to love the austerity of his research site, a day's drive east of the remote central Australian town of Alice Springs. He loved the vivid red colours and the vast, empty proportions of the landscape. He loved jumping into his beat-up old Land-Rover and bumping along the aboriginal trails that took him into the heart of the bush. He mapped alone. And he knew those rocks better than anyone else on the planet did.

But on the field trip in 1993, nobody had called Martin. He was still only a student, and the status-obsessed organizers had arranged everything without involving him at all. It was as if someone brought a field party tramping through your backyard without warning or explanation. Martin was furious. As the trip progressed, he made himself more and more obnoxious. He challenged everything, did his best to humiliate the leaders by pointing out their errors, and one night over dinner he brought one of them close to tears. (Eight years later this highly eminent geologist can still barely bring himself to mention Martin's name.) The more people tried to slap Martin down, the more belligerent he became. Nick was intrigued. Here was someone else who constantly challenged and harried. What's more, he was often right. When Martin argued about the rocks, he did so with little tact but plenty of intelligence. As soon as the trip ended, Nick started talking to Martin about how the two of them could collaborate.

Shortly afterwards, Martin took Nick out to see some other Australian rocks that he'd been working on, just outside Adelaide. Now Nick was the abrasive one. Throughout the trip, he was incessantly challenging and infuriating. He argued every point. "How do you know this rock isn't the same as *that* rock *there?*" By the end of the trip, Martin felt as if he'd been put through a wringer. But he also realized that his mapping had been tested as fully as it could ever be. When Nick is finally satisfied with the picture you paint him, you know it must be right. "Nick's very bloody-minded." Martin says. "He's a contrary person, and that's his value in the world. I've learned a lot from him. I don't speculate any more. I don't just let my lips flap."

Nick and Martin have worked together on and off ever since. They cherish the collaboration. And in annoyance as much about

Paul Hoffman's blowhard style as his scientific substance, they have become firm allies in the anti-Snowball game. Now Martin figured that he had managed to cook up a real alternative to Paul's theory. He was in Death Valley to check out the evidence for this latest challenge to the Snowball.

Daylight had almost gone when the truck finally arrived at the Noonday dolomite, a pale tan slope of rock that marked the cap carbonate overlying the glacial deposit. The rocks were shot through with dark, thin vertical lines that looked like worm burrows, though they had formed long before worms were invented. They were exactly like the tube rocks in Namibia that had so baffled Paul.

And there were other strange structures in the rock. The ancient mud layers were occasionally buckled into chevrons, and the insides of the chevrons were filled with very fine-grained cement. That was confusing. In geology, cement appears when a gap somehow opens in the rock after it has first formed. When the mud was first hardening into rock, something happened to create an internal space, like a crack in the middle of a brick wall. At some time later, fluids pouring through the rock deposited cement, which filled the gap.

But a chevron-shaped gap is strange. To make a chevron, you need to squeeze the rock until you rumple it into a fold. To make space for cement, you need to stretch the rock until a gap opens up. What process on Earth can both squeeze and stretch at the same time?

Martin thought he knew. He jumped back into the truck, eyes shining triumphantly. The signs in the Noonday dolomite were just what he'd been looking for. He was assembling evidence that the carbonate rocks that capped the Snowball deposits came not from some mighty weathering in the inferno that followed the ice, as Dan had suggested, but from an entirely different phenomenon, one that didn't need an inferno, didn't need a total deep-freeze, didn't need any of the things that, to Paul and Dan, made up the essence of a Snowball world. All Martin's model needed was for a strange substance, a chimera, half-ice and half-fire, to be scattered throughout the Snowball world.

Martin's chimera is called methane hydrate and is made of tiny ice cages, with molecules of methane—natural gas—trapped inside. Methane hydrate is amazingly abundant in the world today. Together with other gas hydrates, it harbours more than twice as much carbon as all the known natural gas, oil and coal deposits on Earth. It looks like dirty ice. It's often smelly, too, giving off a whiff of rotten eggs from the sulphurous activity of bacteria that are typically found in the sticky mud alongside. It's very unstable. Today methane hydrate survives only in the frozen Arctic soil, or in the high pressures that exist beneath the sea. Raise the temperature, or bring a lump of hydrate up to the surface, and it will rapidly disintegrate. You can hold it in a gloved hand and watch the ice disappear, fizzing with pops but no crackles. Strike a match, and the gas coming off it will ignite, and then burn with a reddish flame. All that's left behind at the end is a muddy pool of water.[2]

Because methane hydrate is so widespread, many people over the years have championed it as the fuel of the future. The trouble is, the ice cage is so unstable that mining it can trigger disaster. If you accidentally set a batch of methane hydrate decomposing when you're trying to extract it, the methane that bubbles out turns seawater into foam. Because foam is much less dense than liquid water, your drilling ship promptly sinks. In fact, since there's plenty of gas hydrate on the continental rise off the southeastern United States—the western portion of the so-called

Bermuda Triangle—this mechanism has frequently been invoked to explain the mysterious maritime disappearances that are supposed to have taken place there. The idea is that a sudden underwater landslide could reduce the pressure on the hydrate deposit, decomposing it and sending deadly bubbles of gas up to the surface. This is plausible enough geologically, but, sadly, the shipping end doesn't hold up. The Bermuda Triangle simply hasn't swallowed an abnormal number of ships—ask Lloyds of London.

Still, methane hydrates have apparently been responsible for real-life dramas in the past. In the Barents Sea, just off Norway's northeastern tip, hydrate deposits seem to have exploded thousands of years ago, leaving behind giant craters that pockmark the seafloor. That probably happened at the end of the last ice age, when warmer seas destabilized the hydrates there until they erupted like a volcano. And some researchers believe that destabilized hydrates released more than a million cubic kilometres of methane at the end of the Palaeocene, about 55 million years ago.

Martin Kennedy wants to explain the carbonates that blanketed the Earth immediately after the Snowball in the same way. Methane hydrates, he thinks, might have been more widespread than they are today. After all, everyone agrees that conditions were colder then. When warmer temperatures returned, those methane hydrates would have released their gas, to be quickly oxidized and precipitated into the ocean as a blanket of carbonate.[3]

Martin felt that he had found clear evidence for this idea in the Noonday dolomite in Death Valley. That's why he was so excited. The chevrons filled with cement, the rocks that looked as if they had been simultaneously squeezed and stretched, the "worm burrow" structures—all of these could have been caused by decomposing methane hydrates. As the light methane travelled upwards through the heavier mud, it would have created tube-like vertical passageways, just like the dark tubes in the rocks. And if the gas hit an obstruction—a microbial mat, say, which is dense and rubbery—it would have lifted the mat up into a dome, creating both the chevron shape and the cavity that cement would eventually move in to fill. The strange structures in the Noonday dolomite were Martin's missing evidence. "They're stunning," he breathed. "They do exactly what you'd predict. It's better than I could ever have imagined."

Martin wasn't just trying to disprove Paul and Dan. He had other reasons for preferring his explanation of the cap carbonates. Martin, like Nick, is deeply rooted in the world around him. He too thinks that seeing is believing. And he likes his methane theory precisely because it uses processes that we see happening today. There's plenty of methane hydrate in the world right now. This explanation for the caps doesn't require insanely high weathering rates, an extreme hothouse after the ice, a frozen ocean, all the things that make Paul and Dan's Snowball so radically different. "Paul and Dan's Snowball is *really* non-uniformitarian," Martin told me once. "It really worries me when you suddenly evoke a Martian-like world."

But there are some things that Martin is willing to admit, and that Paul and Dan immediately seized on. His methane idea doesn't explain nearly as much as Paul and Dan's theory can. It can't explain the iron formations or the ubiquitous ice. It tells you nothing about why the Snowball might ultimately have ended. And it doesn't necessarily oppose Dan's explanation for the cap carbonates. Even if Martin was right, the hydrates could have been decomposing *at the same time* that acid rain was lashing on to ground-up rocks. There's nothing to stop the two effects working in tandem. Martin's methane-hydrate theory, it turns out, doesn't disprove the Snowball at all.

Martin, however, had another challenge to make. All of Paul's information about the Snowball ocean came from the isotopes in his carbonate rocks, but he only had carbonates from before and after. Martin, on the other hand, had found something extraordinary. He had carbonates from *during* the Snowball. And they seemed to show that Paul's Snowball theory had a fatal flaw.

JUST OVER 600 million years ago, colonies of bacteria floated invisibly in the sea that would one day become northwest Namibia. They hung in the water perhaps a hundred feet below the surface. Undisturbed by wind or waves, they busied themselves with the endless operation of their internal chemical factories. Make food. Consume food. Make food. Consume food. The sea around them turned hazy with the accumulation of their minuscule efforts. As the chemical balance of the water changed in response to their factory effluents, tiny flakes of carbonate sprang out of solution and floated softly down to the seafloor.

Though the Snowball had already gripped the outside world, there was little sign, this deep in the water, of the icebergs passing overhead. Just the occasional dropstone, a boulder released from the ice that would appear suddenly from above, pass gently by the clouds of bacteria, and sink into the flakes of mud that were slowly, steadily accumulating on the seafloor beneath.

Today this carbonate mud has transformed into layers of rock perhaps six inches thick. Its layers stand out clearly among the ice rocks of the Namibian outcrops. They're the colour of rancid butter, occasionally enlivened with a white or tan dropstone. Above and below them the rocks are grey with shattered carbonates, silicates and sand, landslides of debris brought suddenly in from the shore. But the serene yellow layers speak of gentler moments in the life of the Snowball ocean. Made of microscopic carbonate flakes, balled up into tiny spheres called peloids, they are extremely delicate. Intact peloids are a signal that the rock hasn't moved since they formed. If it had tumbled, the peloids would have been smashed to pieces.

The peloidal mud also has occasional cracks, filled with crystals of carbonate cement that jut outwards from the walls of the cracks like the spears of a tiny white picket fence. They too must have formed in the Snowball ocean.

Martin Kennedy had collected samples of these rocks from Namibia back in 1996. He'd also collected similar carbonates from his beloved Australian Outback in the early 1990s. The Australian samples were stromatolites, those strange domed structures created during the Slimeworld, when bacterial mats sat on the growing surface of a rocky edifice, and sediment accumulated beneath them. They were tall, thin columns of carbonate rock, perhaps six inches wide, embedded in yellow dolomite, and they provided another direct window into the Snowball ocean.

Now Martin looked back at these rocks and wondered. Could they help him test the Snowball idea? In early 2001 he returned to Death Valley to collect samples from yet another set of Snowball carbonates, which he'd spotted mixed in with the ice rocks there. These carbonates were called oolites, and were strange rock forms with a caviar-like texture. You find them today in places like the Bahamas. They grow as grains that roll backwards and forwards in the waves, coating themselves with carbonate that precipitates from the seawater. Like the peloids, they're extremely delicate. If you find them intact, they can't have come from some other place, or some other time slice. Since they are mixed in with Death Valley's ice rocks, they too must have formed in the Snowball ocean.

Martin realized that all these carbonate samples gave him a direct window into the Snowball ocean and, with it, a way to test the Snowball hypothesis. Remember that according to Paul and Dan, the Snowball ocean was essentially lifeless. That was Paul's first idea, the one that set the Snowball rolling. Paul had worked it out by looking at the ratio of light and heavy isotopes in carbonate rocks from just before the Snowball time. Heavy carbon equals life. Light carbon equals no life. And immediately before the Snowball, Paul had found unusually light carbon in his carbonate rocks. He concluded that many living things must have died off as the ice advanced, leaving only a few small groups to huddle together and wait for the thaw.

But because Paul had never managed to find carbonates from the Snowball ocean itself, he had no direct evidence for how much life there was then. If he was right, and the Snowball was an intensely cold, barren time, there was scarcely any life around. In that case, carbonates formed chemically from the Snowball ocean itself should have been light. If he was wrong, if the ocean was full of life, the carbonates from the seawater of the time should be heavy.

All right, then, Martin thought to himself. There's a hypothesis. Let's test it. He set about measuring the isotopes in all of these rock samples. And every time, he found the same thing.

They were all heavy.

Life in the Snowball ocean was apparently flourishing. There couldn't have been the extreme freeze-over that Paul demanded. Hot news—the Snowball wasn't that cold after all. This looked like the fatal flaw in Paul and Dan's hypothesis. Martin hastily wrote up the results and submitted them to a journal called *Geology*.

Academic journals decide what to publish on the basis of "peer review", where anonymous scientists say what they think of the work. And one of the scientists to whom the editors at *Geology* sent the paper was Paul Hoffman. Paul's review was savage. He waived his anonymity ("I always sign my reviews") and firmly recommended that the paper be rejected. It contained, he said, a basic geological error.

According to Paul, Martin's samples had nothing to do with the Snowball ocean. They were, Paul believed, simply broken-off pieces of older rock. That would destroy Martin's argument. If the carbonate cements were from a much older time, their isotopes would be perfectly acceptable. Of course the ocean that existed long before the Snowball was full of life! The problems for life came only with the ice. If Paul was right, Martin was embarrassingly wrong.

Paul's review was still in the post, on its way to *Geology* and thence to Martin, when the two of them arrived in Edinburgh in June 2001 for a workshop about the Snowball. Paul was in pugnacious mood. A few days earlier he had stabbed his finger on the offending picture in the draft of Martin's *Geology* paper. "I'm hoping he'll show this photograph," he'd told me. "I'd like it to be demonstrated in public that this guy is incompetent in geology."

But Dan Schrag, Paul's smoother of relationships, was also in Edinburgh, and had decided to try bringing Martin on board. He took Martin to a pub with a crowd of other conference participants, and soon they were chatting with great good humour. They were discussing science, Snowballs, Dan's recent flight over Edinburgh in a friend's microlight aircraft. Martin was saying that he didn't want to be part of an "anti-Snowball team", that he didn't consider himself on anybody's "side", and that Nick should never have done the "snow job" talk, because it had just polarized everyone's opinions to little effect. What Martin and Dan both wanted, they agreed, was to keep talking, test the waters, come to the right answer.

Paul was in the pub too, struggling to keep quiet, doing his best not to spoil Dan's efforts. "This business of Martin Kennedy trying to kill the Snowball with his cements," he said to me later. "Half of me wanted to expose him and discredit him, so that nobody will believe his 'facts'. But half of me is appalled that I'd want to do that. It's not that I have any warm feelings for Martin. He's been a thorn in many of our sides for years. But to try to humiliate him in public would be cruel. Honestly, I'm not a malicious person. I'm certainly capable of being malicious, but it tends to be when I haven't thought it through."

Still, Paul couldn't quite contain himself. When Martin stood up to leave, Paul stood up too, determined to assert his conviction that the contentious cements were from older rocks. "You showed those cements and said they'd formed in place!" he said loudly, looking directly at Martin. "But I know what they are." All conversation around the table ceased. Everyone was staring. "I feel like I always did when my parents were fighting," interjected one of the students, *sotto voce*. "Poor Martin," mouthed another, silently. Dan sighed. He looked at Martin and said carefully: "Paul has sent you a signed review making that comment . . ."

"Yes," Paul interrupted, louder still. "And I had to tell you about it today in private or tomorrow in public." Martin by now was looking aghast. Dan jumped up and headed towards him. "Look, it's OK", he said soothingly. "You'll talk about it tomorrow." He put his arm under Martin's elbow and guided him out of the pub.

The next day, at the Snowball workshop, Paul was doing his best to be friendly again. Every so often during the presentations,

Paul would turn and mutter something conspiratorially to Martin, who would grin and whisper back. And at the end of the day, as the assembled researchers were leaving the lecture theatre, Paul walked over to Martin and shook his hand firmly.

"All the best, Martin," he said. "And I'm glad you're back . . . in the publishing world." Martin had spent some time working at Exxon, where commercial research is conducted behind closed doors. He'd only recently rejoined the academic world and started being allowed to publish his research again. Paul was trying to be polite. But as soon as he said it, he and Martin both thought of the cements manuscript, the one Paul had reviewed unfavourably, the one that might well not be published now because of Paul's comments.

"I guess I could have put that better," Paul said uncomfortably. Martin shrugged. "Well, you know . . ." he said, and turned to leave.

Paul started to follow him up the stairs. "But it could still be accepted, right? Mine was only one review."

"I think you have more influence than you realize," Martin replied.

"I was trying to spare you embarrassment."

"Oh, I don't think you have," said Martin. "In fact, you're going to spare me tenure if this goes on." Unlike Paul, Martin was not yet a tenured professor. Whether his position at the University of California at Riverside was made permanent would ultimately depend on a careful assessment of what papers he had published and how successful his research had been.

Paul was still trying to justify his review. "I wanted to show you these pictures," he said. "Listen, have you got time now?" Martin hesitated, then shrugged again. "OK," he said.

The lecture theatre was empty now. Paul climbed quickly to the top of the stairs, and began fumbling with his slides. Click. He showed a grainy image of a Namibian carbonate rock. This, he said, was the older rock formation, the one he believed Martin's samples originally came from. Click. There was another. This was the younger Snowball formation, with a chunk of the older rock embedded in it. Martin stared at the screen. He looked appalled. "Paul," he said, "that's not what I collected."

There was an awkward silence. Paul was standing stock-still at the top of the stairs. "What I wanted to say was . . ." he began, but Martin interrupted. He was clearly upset now, fighting to keep control of his voice. The images had convinced him that Paul's damning review was based on rocks that were entirely different from his own. "The cements that you showed are not the same as the ones I collected, Paul," he said formally. "But thank you for showing me the photographs." And then he turned abruptly and left the room.

A few weeks later, Martin's paper was accepted after all.[4] Even if his data hadn't impressed Paul, the other reviewers found his analysis sufficiently compelling. Paul, it turned out, had concluded his damning review with one of his favorite quotes: "False facts are highly injurious to the progress of science for they often endure long: but false views, if supported by some evidence, do little harm, for everyone takes a salutary pleasure in proving their falseness."

This comes from Charles Darwin, the originator of the theory of evolution, and it makes an important point. When people argue about ideas—"views", in Darwin's words—all the arguments have to be based on the available "facts". If one of the facts is wrong, the whole edifice can crumble—taking *all* the ideas with it.

But whose "facts" were wrong—Paul's or Martin's? Had the

rocks come from the Snowball ocean or not? Paul might be right about the "picket fence" crystals from Namibia and their related cements. It is possible that the crack they grew in was from some earlier time, and that the whole lump of rock had then broken off and been amalgamated into the younger Snowball formation. In the end, Martin left those samples out of his paper, just in case. But the peloids are harder to argue away, and the caviar-like oolites harder still. To break those off from older rock and transport them and mix them into a new formation would surely have smashed their delicate structures. They really did seem to come directly from the Snowball ocean, just as Martin said.

So was this indeed a fatal flaw in the Snowball argument? After all, the carbonates leading up to the Snowball had been light, and Paul had assumed that the ones that formed during the Snowball would be the same. But he hadn't made any direct assertions about this in his papers. Unlike Martin, Paul had no carbonate rocks from the Snowball period; with no data to interpret, neither he nor Dan had thought particularly hard about what the isotopes would be like. Now, though, they had an incentive. Spurred on by Martin's findings, Paul and Dan focused on this issue. And for two different though complementary reasons, they realized that you'd actually *expect* the Snowball ocean to be heavy—just as Martin had found. Ironically, Martin's heavy rocks didn't conflict with Paul and Dan's model at all.

Paul had realized that the oceans would have been heavy because they contained old material, dissolved from rocks on the seafloor. The floor of the Snowball ocean was lined with carbonate rocks that had been created when life was still abundant. And the acidic seawater would have dissolved this old carbonate, just as acidic lime juice can dissolve a marble cutting board. The effect, says Paul, was to change the signature of the ocean. If you pour hot milk onto cocoa powder, the milk turns brown because the dissolved cocoa swamps the milk's original colour. Similarly, the "heavy" signals of plentiful life dissolved from the old seafloor carbonate would have swamped the "light" signal from the largely lifeless Snowball ocean.

Dan has discovered another effect that reinforces this one. His answer involves a more arcane issue, beloved of geochemists and understood by few others. The carbon isotopes in the ocean don't depend only on the activity of living things; they are also affected by the way carbon dioxide gas migrates from the atmosphere into the ocean. And this in turn depends on what proportion of carbon the atmosphere already contains.

Nowadays the atmosphere has only a tiny proportion of carbon, less than 5 per cent. But the Snowball was very different. According to Paul and Dan's model, carbon dioxide gas had been building up in the atmosphere for millions of years. The Snowball atmosphere contained a much higher proportion of carbon, and that would have made all the difference.

Dan did the sums. He crunched through all the equations that predict how this change would affect the ocean isotopes. And he came up with a number that matched—precisely—the heavy values Martin had found. Even if the Snowball ocean was totally lifeless, the carbonate cements would be just as Martin had measured.[5] This was no fatal flaw. Paul and Dan concluded that Martin's evidence *confirmed* the Snowball theory.

WHEN NICK Christie-Blick, Martin's co-author on the cement paper, heard Paul and Dan's new arguments, he immediately cried foul. Paul and Dan, he said, were simply shifting their goalposts.

How could he and his fellow critics test their theory if they kept changing what it said? Paul and Dan responded that they were naturally modifying a young theory, making it richer and fuller.

Who's right? Well, science works at its best when somebody puts forward a theory and everyone else tries to pull it down. Sacrosanct scientific philosophy holds that no theory can ever be *proved.* A theory can only be *disproved,* and the longer it survives the attacks against it, the more confidence you can place in it—while never knowing for certain if it is right. Following the philosopher of science Thomas Kuhn, many see science as a procession of revolutions, where a prevailing paradigm holds sway in researchers' minds until it is finally disproved and a new one takes its place.

The trouble comes in this process of disproving. Scientists often—perhaps usually—find it hard to let go of a theory that they care about. When some devastating new finding shows it to be wrong, that's hard for them to accept. There are many theories whose proponents have clung on to them for too long, rendering them more and more elaborate in a desperate attempt to accommodate the findings that disprove them and ward off the inevitable end.

But science moves much more in fits and starts than a simple reading of Kuhn's paradigm shifts would suggest. And it can be hard to decide whether a theory has truly been disproved. Often counter-arguments can be incorporated into the theory itself until it becomes richer for having adapted and allowed itself to grow. If a theory comes under attack when it's too young and raw to defend itself, it can also be destroyed prematurely. Wegener's theory of continental drift is one example of this. And Alvarez's asteroid hypothesis about the death of the dinosaurs could have encountered the same fate if someone hadn't found the crater—the "smoking gun". Even that was lucky. The crater could have long since been swallowed back up into the Earth's interior, as much of the rest of the Earth's crust has been since the time of the dinosaurs. Alvarez's theory might have been right, and yet could still have been killed.

Though Martin's cement paper may not have tripped up the Snowball idea, as he had thought it would, it had taught Paul one lesson at least: when under attack, try to spread out the target area. The next time I saw Paul at the beginning of a new lecture tour, the Snowball had become "Kirschvink's theory". Every time he mentioned the idea, Paul was reminding his audience that it had come, in much of its modern incarnation, from the insights of Joe Kirschvink, the Caltech professor who had dreamt up the volcanic aftermath, and coined the name Snowball Earth.

Paul even described himself as "Kirschvink's bulldog". That was neat, a direct reference to Thomas Henry Huxley, who earned the nickname "Darwin's bulldog" for his blunt and ferocious defence of Darwin's ideas on evolution. Darwin shrank from crusading against the disapproving Anglican establishment on behalf of his worrisome new theory. Huxley, however, had no such qualms. At a debate in 1860, when Bishop Samuel Wilberforce asked sarcastically whether Huxley would prefer to be descended from an ape on his grandfather's or grandmother's side, Huxley reportedly replied thus:

> If the question is put to me, would I rather have a miserable ape for a grandfather, or a man highly endowed by nature and possessed of great means and influence, and yet who employs these faculties and that influence for the

mere purpose of introducing ridicule into a grave scientific discussion, I unhesitatingly affirm my preference for the ape.[6]

Meanwhile, Martin's paper on the cements had done something else for the Snowball idea. He had finally focused attention back on to the question of what exactly was alive in the Snowball ocean. This was something that was beginning to worry many biologists. They knew that certain creatures must have survived the Snowball: bacteria, of course, in their enveloping mats of slime; slightly more sophisticated—but still single-celled—creatures, with their internal chemicals neatly packaged rather than floating freely in a soup; simple algae, brown and green and red. All these creatures left their faint fossil traces in rocks from both before and after the ice, so they, at least, must have lived through it. But the scale of Paul's freeze-over was troubling. If it was as severe as Paul insisted, how could *anything* have remained alive at all? This was to become the Snowball's next test.

ICE IS an extraordinary substance. Subtle shifts in its structure can render it white or green or blue, translucent or opaque. It can shatter like glass, or creep like treacle. Ice is a tough building material, as strong as concrete. In the Second World War, plans were even developed to create giant aircraft carriers called "bergships" out of ice. They might have been built, too, if the range of aircraft hadn't increased enough to render them obsolete. Russian empresses used ice to build vast, glittering palaces: "The delightful material gave a new, fantastic beauty to every feature, sometimes white and sometimes clear green—dark and opaque where the shadows fell, and almost transparent in the sun. No dream castle of jasper or beryl . . . could be more beautiful than these wonderful buildings of ice."[7]

Ice is alien to life. Part of the attraction of the Antarctic ice cap is that the essentials of life—food, water, fuel and shelter—have all been stripped away. Explorers have talked for decades about the sublimity and purity of this landscape. "During the long hours of steady tramping across the trackless snow-fields, one's thoughts flow in a clear . . . stream," Antarctic explorer Sir Douglas Mawson wrote, while trying to explain his urge to return. "The mind is unruffled and composed and the passion of a great venture springing suddenly before the imagination is sobered by the calmness of pure reason."[8]

But there is danger as well as purity in this escape from life. Remember Wegener in Greenland, Scott in Antarctica, and Hornby in the bitter winter of the Canadian Arctic. Ice also kills. Every cell in your body is a squashy bag of water, with just a few other chemicals thrown in. If this water freezes, jagged crystals of ice appear, and they slash and tear at the cell's fragile walls. These membranes also spring leaks when their molecules begin to congeal together into clumps, like fat cooling in a frying pan. Within the cell, proteins unwind their complicated loops and become flaccid. With great care, and clever technology, certain cells can be preserved on ice—sperm, eggs or bone marrow. But for the most part, life depends on water; and ice brings death.[9]

That's why the biologists were so worried by Paul Hoffman's Snowball. If ice covered the world, how could even single-celled life have survived?

Water wasn't the whole problem. The oceans in Paul's world wouldn't have frozen solid, largely thanks to another strange property of ice. Most solids don't float. They become denser when they freeze, and in a bath of their own liquid, they'll sink. But ice is the exact opposite. When water turns to ice, its molecules become more loosely bound, forming a lacy network that's full of space. That's why ice floats, and why the Snowball oceans didn't freeze completely. If icebergs sank, lakes and oceans would freeze from the bottom up, instead of just growing an ice skin on their surface. So, even as Paul envisioned it, there would still have been plenty of liquid water *within* the Snowball oceans.

But living things also need sunlight. Because Paul argued that the entire surface of the ocean was frozen over, all the water would be beneath that ice layer, blocked from the sun. And for most of the simple denizens of the Snowball, darkness would mean death. Even the creatures that didn't make their living using sunlight depended on the ones that did. Living things needed both liquid water and sunlight, and for that, the ice had to have holes.

At first, Paul wouldn't hear of this. The oceans were fully covered with ice, and that was final. But once again he was forced to change his mind. And the impetus for this came from a rival ice world, a pretender to the Snowball crown, which began to tug at everyone's attention. This was a newer, gentler snowball, with a moniker of its own: "Slushball Earth". Climate modellers created it. They use computer programs to do what geologists can't— rerun the Earth's experiment and see what happens. As soon as they heard about the Snowball, they fired up their machines and tried to make one.

They couldn't.

However much the modellers wanted to generate an ice-covered world, their computers wouldn't oblige. Modelling had developed into a much more sophisticated affair since Mikhail Budyko's primitive attempts first turned up the "ice catastrophe" back in the 1960s. And the modern models stuck at a sort of halfway house, where ice advanced to somewhere near the tropics, but no further. A few models could generate ice *on land* near the equator—which would explain the ice rocks from Australia. But the equatorial *oceans* remained stubbornly ice-free.[10]

So the modelers began to talk of an alternative to Paul's "hard" Snowball, a new, softer variant. Nick Christie-Blick thought this Slushball was a wonderfully moderate solution to the Snowball conundrum; neither one extreme nor the other, it was a comfortable answer. It could also explain some of the evidence that had consistently bothered Nick. For instance, in many parts of the world the ice rocks are hundreds of feet thick. To make those deposits, icebergs had to be free to wander offshore and melt and drop their loads on the seafloor, and they certainly couldn't do this if the oceans were fully frozen. Paul argued that the ice rocks formed at the beginning and the end of the Snowball, when there was still a little open water. But Nick felt that to make such thick deposits, the process would have to continue throughout the Snowball. Open oceans at the equator, he felt, provided the perfect answer. For the biologists, too, the Slushball was just right. This, they felt, was exactly what life needed.

Paul, however, hated the Slushball. He called it "Loophole Earth", and said the models needed a few reality checks of their own. The Earth's climate is insanely complicated, and nobody claims that its every nuance can be encapsulated inside a computer. Modellers are good at reproducing today's climate mainly because they can compare their model output with records of temperature, wind and weather. But the only Precambrian weather reports are the ones written in rocks. And according to Paul, the Slushball came nowhere near explaining this geological evidence.

SNOWBRAWLS

It couldn't account for the ironstones, the cap carbonates, or the strange chemical signatures in the rocks. Most important of all, it couldn't explain the extremely long *duration* of the ice.

Paul pointed in particular to the findings of Nick Christie-Blick's graduate student, Linda Sohl. Her magnetic work had shown that the glaciations must have lasted at least hundreds of thousands, if not millions, of years. The Slushball, said Paul, simply couldn't last that long. It was precarious, like a pencil balanced on its tip. Nudge the model world one way or another, and you would quickly force it to choose: Snowball or no-ball.

If you cooled the Slushball a little, Paul said, ice would quickly take over. White ice reflects sunlight, which cools the Earth, which breeds more ice in a runaway cycle, which, Paul said, would freeze over the tropical ocean. Warm the Slushball a little, on the other hand, and its ice would soon vanish. Warming melts ice, which opens up dark patches of ocean, which absorb more sunlight until all the remaining ice races back to the poles.

What, then, was Paul's explanation for how life survived the ice? Well, living things are extraordinarily resilient, especially simple ones. Bacteria survive—somehow—at the South Pole. Other bacteria have shown up beneath glaciers, and even inside solid rock. Unknown to the authorities, a small colony of *Streptococcus mitis* hitched a ride to the Moon in 1967 inside an Apollo TV camera, and the bacteria were still alive three years later when the camera was brought back to Earth. They had managed to survive without food, water or even air. Hot springs are often brilliant with living colour. The steaming, acidic pools of Yellowstone National Park, for instance, contain vivid bacterial patches of orange, red and green despite their boiling temperatures. Life has a habit of finding its way, no matter what.

The biologists pointed out, however, that many of these resilient creatures are weirdly adapted to their extreme conditions, whereas most of the ones that survived the Snowball were apparently more normal in their requirements, particularly in their need for sunlight. There had to be sunlight. There had, the biologists said firmly, to be holes.

So Paul and Dan changed tack. They obviously needed to provide some refuges for life within the frozen seas. What kind of openings might there have been? Well, any hot spring or volcano on a shallow enough ocean floor would have created at least a small hole in the ice above it. Also, the Snowball was not uniformly cold. Though global temperatures would initially have plunged to around minus 40 degrees C, they would gradually have risen as carbon dioxide built up in the air. And the equator would always be warmer than this bleak global average. Soon the ice at the equator would grow thinner, perhaps even thin enough to crack periodically.

Thinking the question through further, Paul and Dan also realized that the Snowball's stronger seasons would also have helped living things cope. Even if winter temperatures near the equator were 30 degrees below zero, summers could have crept above freezing for a few days each year. In melted puddles and ice cracks, living things could then have grabbed their chance to make and store food, as they do in Antarctica today. And even in winter there could well have been other patches of open water among the ice. In today's frozen oceans, odd currents keep certain places—called polynyas—ice-free throughout the year. Whales trapped in the pack ice use these open patches as breathing holes while they wait for spring to return and release them.

For the biologists, this line of thinking was much more encouraging. But were there *enough* refuges? Could each individual species huddle together in a big enough group to survive until the Snowball finally melted?

To find out, Dan Schrag called a friend, another hot young scientist, Doug Erwin, from the National Museum of Natural History in Washington, D.C. Doug is an expert on ancient life, and he also knows a fair amount about ecology in the modern world. To protect an endangered species, Doug said, you have to maintain its genetic variety. The genetic material that passes from one generation to the next is constantly changing—and not always for the better. In an isolated group—a herd of elephants in a national park, say—dangerous mutations can spread quickly. For the species as a whole to survive, any one group must contain enough individuals, enough variety, to dilute this danger. And there must be enough separate groups that if a few of them fail, the rest will still pull through.

Doug realized that the same would apply to the Snowball's inhabitants. He made a list of all the different species that needed to make it through the Snowball. Then he used conservation models to calculate two numbers: how many individuals of each Snowball species you'd need in a given refuge, and how many refuges you'd need overall.

The answer astonished both Doug and Dan. It was far easier than they'd expected. To get virtually all of the species through the Snowball, you only needed something like one thousand different refuges. And each refuge only needed to house around one thousand individuals. What's more, the Snowball creatures were no elephants. "Do you know how much open water you'd have needed to support one thousand of these individuals?" Dan demanded of me as we sat in a café. "This much." He spread his hands apart until they outlined a region of air the size and shape of a dinner plate.

For Doug and Dan, at least, this solved the problem of

survival. Of course you could poke one thousand small holes into the Snowball ocean. You could probably make tens of thousands without compromising the model, and many of them could be much bigger than a dinner plate. With this many refuges, says Doug, every species that needed to make it through the Snowball could do so easily. There's no need for a Slushball. Life could survive an all-out, full-on Snowball with no problem at all.[11]

SO FAR the Snowball theory has survived every challenge that's been thrown at it. And thanks to Paul and his proselytizing, there has now been a radical transformation in scientific attitudes to the ice rocks. In a few short years Paul has achieved the feat that eluded Brian Harland, Joe Kirschvink, and anyone else who became intrigued by the Earth's Precambrian ice rocks. Virtually everyone now says that this was a time of extraordinary ice, cold and catastrophe. Even critics like Nick Christie-Blick, who still believe in the Slushball, admit that ice went almost all the way. Paul has taken an idea that was once too shocking to be considered, and brought it into the scientific limelight.

He could have stopped there. But there's another part of the Snowball story that he'd dearly love to be true. While Paul continues to strengthen his geological case, he's also fascinated by the biological implications. Was the Snowball the creative spark for the new life that followed?

Paul has believed from the beginning that the ice and its aftermath somehow triggered life's biggest evolutionary moment since it first appeared on Earth: the switch from simple to complex. Without the Snowball, he thinks, there would have been no animals, no richly diverse Earth, and no people to argue over it. Paul, though, is not a biologist. What do the experts say?

NINE

CREATION

Billions of years had passed, nearly nine-tenths of Earth history, when life finally made its vital leap into complexity. Now at last it could move on from dull primordial slime, and begin inventing the fabulous life-forms that we see today.

Of all the innovations conjured up by evolution, this was the most dramatic. It was the world's first industrial revolution. Before then, each individual cell had to be master of all trades: eat, digest; excrete; reproduce; perform all the essentials of life within one small squashy sac. Afterward, mighty corporations of cells sprang up to share the load. Specialization became the rule. Thanks to structural cells, bodies could grow large and adopt inventive new architectures. Muscle cells could move these bodies to new grazing grounds. Sensory cells could warn of danger, appendage cells could rake in supplies. Cells evolved to regulate temperature, transport information, innovate and consolidate.

And this specialization opened up a world of possibilities. Suddenly, in the Earth's late middle age, life began frantically procreating, evolving and developing new forms. First came trilobites and ammonites, then dinosaurs and octopuses, dromedaries, whales and wallabies, as the new complex creatures competed to find ever more imaginative ways of exploiting the world's resources. Life as big business was wildly successful.

Then why did it take so long? Though the history of life is ambiguous, traced through an imperfect record of fossils and rocks, most researchers believe that complexity was invented somewhere between 550 and 590 million years ago. That's after more than 3 billion years of simple, single-celled slime.

Biologists have been trying for decades to understand why complex life appeared on Earth at that particular moment. And then, along came Paul Hoffman, talking of global catastrophe. Paul's evidence suggested that at least two, and possibly as many as five, successive Snowballs had rocked the Earth starting around 750 million years ago. Most significantly, this series of Snowballs ended 590 million years ago, just around the time complex life was beginning to emerge. The news of Paul's ice sent biologists racing back to their fossils. "What did this?" they started asking themselves. "Was it the Snowball?"

These are early days for the biological part of the Snowball theory. Some biologists are automatically dismissive of *any* idea that makes biology subordinate to geology. "Genes don't care about the weather," one researcher told me. "Adding ice cubes doesn't give any explanatory insight," e-mailed another.

But others are intrigued by Paul's findings. There are now

signs that complexity really did appear soon after the ice receded. And though the picture is still far from clear, many biologists are beginning to think the Snowball could indeed have been the trigger.

To FIND the cause of a historical event, you first need to know when to look. And until recently, most biologists have assumed that complexity arose with an event called the Cambrian explosion. This episode has grabbed all the early-life attention for decades.

EVOLUTION'S BIG BANG! screamed the front cover of *Time* magazine on 4 December 1995. "New discoveries show that life as we know it began in an amazing biological frenzy that changed our planet overnight." The animals that reared up on its cover were from the beginning of the Cambrian period, around 545 million years ago. At first sight, that seems like a serious problem for Paul. The Cambrian explosion can't possibly have been triggered by his Snowballs. They ended around 590 million years ago,[1] and 45 million years is far too long to sit around with a lighted fuse waiting for the bang. Even Paul admits this.

But he also says that the Cambrian doesn't deserve quite as much attention as it receives. The beginning of the Cambrian was certainly a burgeoning, inventive time for life. During this rapid burst of new evolutionary shapes and strategies, the foundations were set for every modern family of animals. The Cambrian fossils have been known for centuries; they mark the end of the Dark Ages without fossils and the beginning of geological and biological enlightenment. They are Stephen Jay Gould's "Wonderful Life".[2] But all this fame has come to them mainly because they were *easy to preserve*. They show up everywhere. At the beginning of the Cambrian, life invented skeletons: scales, shells, spines, all the sorts of bodily supports that stick around long enough after death to turn into clear, unambiguous fossils.

So the Cambrian fossils weren't the first complex animals, any more than language began with the printing press, or with papyrus. Complex life could easily have been around for millions of years before then, and just not left such a clear record in the rocks.

The invention of multicellularity was certainly a *prerequisite* for the Cambrian explosion. Some biologists even say that it made the Cambrian explosion inevitable. Of course, life began experimenting with its new toy, exploring the many new possibilities it now had for shapes and functions, tissues and organs. With complexity already in place, the Cambrian explosion was just regular evolution in action.[3]

So forget the brash fossils of the Cambrian. To find the real moment that life learned to use many cells instead of one, biologists need to seek out creatures that are much more mysterious. If Paul Hoffman is right, and the Snowball truly triggered the invention of complexity, the world's first complex creations must have appeared shortly after the ice receded. The question is, did they?

CRUNCH! JIM Gehling's foot lands on a coke can and squashes it into the ground. He picks up the can, points to the misshapen circle its trace has left in the mud, and grins. "Go on," he says. "Look at that, and tell me what shape the can was originally, or what it was used for."

This is one of Jim's favourite metaphors for the work he does: reconstructing some of the first new creatures to emerge after the end of the Snowball. His task is extraordinarily hard. At least people studying more recent fossils such as dinosaurs have something concrete to dig up—bones, scales or shells. But the creatures that Jim studies had none of these attributes. They lived before skeletons had been invented. Their bodies were soft, like jellyfish. And the fossils they left behind are like the smudgy circle from the coke can—indistinct impressions, squashed into ancient mud. From these scant clues Jim and his colleagues have been trying to figure out whether these jelly-creatures were the world's first complex animals.

Jim is from Adelaide, Australia. He is in his mid-fifties, tall and slim, with white spiky hair, a long, thin face and a centurion's hooked nose. His eyes are deep blue, and—consciously or not—he tends to wear shirts that highlight them perfectly. His smile is engaging, his manner easy. He delights in his fossils and—most unusually, in the jealous field of palaeontology—he loves to share them. He is a prime candidate for the world's nicest man.

Everybody likes Jim. They all say so repeatedly, even Paul Hoffman, though Jim recalls that he and Paul fought heatedly about something or other when they first met. Jim has the knack. He can disagree with, and even correct, the most egotistical brains in the business without causing any apparent offence. He's not like Paul. You don't want to impress him, or struggle to win his approval. But within a few hours of meeting Jim, you find yourself confiding in him. That's why so many people claim him as their best friend.

Jim first encountered the jelly-fossils as an undergraduate at the University of Adelaide, working for a pioneering palaeontologist, Mary Wade. Mary was an eccentric but enthusiastic teacher. She and Jim camped among the fossils, and—this was the

1960s—she brought her eighty-year-old mother along as chaperon. From Mary, Jim caught the bug. He became addicted to hunting for new fossils, and trying to make sense of the ones he'd already found. What shape was it? How did it live? How did it die? He stayed on for a master's degree, but he quickly realized that job prospects in the field were poor. Jim didn't want to leave Adelaide—his family was settled there—and the only slot for a palaeontologist at the university was already taken.

So he started work teaching general science at a local teacher training college. Though he loved teaching and his students loved him, he couldn't stop thinking about fossils. He'd go to the library and find himself drifting toward the palaeontology journals rather than the ones he was supposed to be looking up. He read about the latest fossil research each night, and spent every holiday out in the field or attending fossil conferences. He produced so many academic papers that many palaeontologists were astonished to discover he was a part-timer.

Finally, when the kids had grown up and left home, Jim quit his day job to concentrate on fossils full-time. His friends were delighted. Bruce Runnegar—a brittle Australian professor at UCLA, with a wicked grin and an acerbic sense of humour—immediately invited Jim to come to California and complete the formality of a Ph.D. (He had already published more research than all the other graduate students put together.) Another friend, a bright-eyed Canadian palaeontologist named Guy Narbonne, arranged for Jim to spend time studying fossils in Newfoundland. Everyone liked Jim in Newfoundland, too. When they heard he was in town, people would appear with gifts for him: a cake, or a bag of berries. He was embarrassed by this.

Now Jim has a precarious adjunct position at the South

Australian Museum in Adelaide. It brings him an office but little money, and his duties as an exhibition organizer take precious time away from his research. There are still no openings at the University of Adelaide, and Jim is struggling to go on with his fossil work. He is neither angry nor bitter about this. He is the most well-adjusted person I have ever met.

Jim's fossils come from a place that has figured many times already in the Snowball story: the Flinders Ranges of South Australia. In 1947, geologist Reg Sprigg spotted what looked like squashed, petrified jellyfish in the rocks of an abandoned mine near Ediacara Hills, on the western edge of the Flinders. Similar creatures have since been found in rocks around the world,[4] but they are still collectively called Ediacarans in honour of Sprigg's discovery.

There's little point in visiting Ediacara today. These fossils are worth tidy money on the open market, and the site has been thoroughly despoiled. Any samples that weren't removed in daylight by palaeontologists were stolen in the night by people using crowbars and mechanical diggers. There's nothing left to see.

But if you swear not to reveal its whereabouts, Jim can take you to a secret location where the fossils are still intact, and in place. You go in the early morning or late afternoon to catch the slanting rays of what Jim calls "fossil light". First you take a paved road north into the Flinders, then a bush track that's strewn perilously with rocks. (Don't offer to drive; Jim doesn't like being driven. But he's gracious about it, and he's also such a good driver in the bush that you probably won't mind.) The landscape is muted, as if faded by the sun to shades of pale terracotta and drab olive grey. On the left, a line of straggly gum trees marks the bed of a dried creek. The ground around is stony, a desert pavement scattered with squat, round saltbushes. Apart from the wedge-tailed eagle sheltering in one of the gums, and the ubiquitous, irritating Aussie flies around your face, there is no life to be seen.

The track swings around a corner and stops at the foot of a gentle hillside, covered with slabs of pale stone. They are irregularly shaped, an inch or two thick and several feet across like broken, prophetic tablets. Jim climbs to one of them, turns it over, and begins to scrub off the dirt with a yellow brush that he has pulled out of his pack. ("Here's the main instrument for this sort of work. A nylon dishwashing brush. Two dollars.") And then he holds the slab out for inspection.

The reason for going early in the morning is now clear. At first you see only the stippled red underside of the rock. But then the slanting light casts shadows that resolve into an oval indented image like a giant thumbprint, perhaps six inches long. The creature that left this imprint is called *Dickinsonia,* one of the icons of the Ediacaran world. Its body is segmented like a worm's, and split by a groove running down its centre. Perhaps this was a stiffening rod for its soft body. Perhaps it is the trace of a gut. At one end, the strange parabolic segments are slightly thicker and wider than at the other. Unlike the inhabitants of Slimeworld, this creature knew the difference between head and tail.

Now Jim is turning over more slabs, and finding more fossils. There's a squashed, sponge-like creature called *Palaeophragnodictya,* revealed as a small disc set slightly off-centre within a larger one. There's another disc with a set of grooves inside it, like the outline of a cartoon arrowhead. *"Aspidella,"* Jim says, and then moves on. Some Ediacaran fossils have flouncy, frilled edges like a Victorian petticoat. Others have discs and stems and branches. One looks like a Roman coin, another like a sheriff's badge—a tiny, five-pointed star inside a ridged circle. Some are truly enormous. One *Dickinsonia* found at this site, Jim says, was more than three feet long.

There's something extraordinary about seeing these ancient ancestors lying in front of you, everywhere you look. Perhaps one of these slabs bears a single indented shape that will shock the biological world. There might be a new species, one that you could name after yourself. The spirit of the hunt catches you, and you start to lift slab after slab. You find more ghostly *Dickinsonia* shapes; then a *Spriggina,* with a long, ridged body and blunt head. But then, suddenly, the sun is too high in the sky, and the images vanish.

The Ediacarans imprinted on these rocks lived—and died—in a shallow, sandy seafloor close to shore. They were the first large creatures to appear on Earth after the long epoch of microscopic slime. Theirs was an innocent age. Predators had not yet been invented, and big, defenceless sheets of flesh like *Dickinsonia* could lie around on the seafloor with impunity. "If you want to have a Garden of Eden from some time in the history of life, this was it," Jim says.

But death still came to these particular unfortunates, in the shape of a storm that stirred up the peaceful sea and brought sand cascading down to smother them where they lay. Each thin slab of rock on the hillside was created during one such underwater sandstorm. Even then, the Ediacarans' soft bodies would have rotted away to nothing, if Slimeworld hadn't intervened to preserve them. Most of the fossil slabs have a rough, stippled texture like elephant skin, the remnant of the slimy bacterial mats that were draped over the Ediacaran seafloor. A few years ago Jim realized that these mats also helped preserve a record of the Ediacaran fossils.

The idea came from an old image. As a child, Jim was flicking through an encyclopedia when he saw a picture of the death mask taken from the corpse of the notorious outlaw of the Australian Outback, Ned Kelly. Jim can still remember the imprint that Kelly's eyelashes had left behind, and the shape of his chin. And when Jim was studying the fossil slabs, he realized that the slimy mats would have provided each Ediacaran with a death mask just like Kelly's.

As soon as the Ediacarans died, bacteria would have rushed to cover them, greedily absorbing their nutrients, and extruding chemicals that bound the sand above into a tough, yellow mineral of iron called pyrite. Even when the soft body rotted away, this pyrite shell would have stood firm. Now, hundreds of millions of years later, that same iron crust survives on the underside of each sandstone slab; it has rusted now to red iron oxide, but still provides a faithful mold of the creature that once lay beneath.[5]

This means of preservation captured not just individuals, but a slice of life, a snapshot of the Ediacaran seafloor. Unfortunately the sand also squashed many of its victims before their death masks formed. This has left Jim and his colleagues with a quandary. Some Ediacarans were born flat, some became flat; how do you tell the difference? Interpreting the impressions of these crushed bodies is certainly an art—hence the business earlier with the coke can. ("Arm-wrestle for it!" one exasperated onlooker told the researchers who had spent most of a field trip wrangling pointlessly about the appearance—when alive—of a particularly smudgy fossil.) But there are clues in the rocks, if you know how to read them. Are the fossils ever folded over? Then they must have been flat when they were alive. Are they all lying in the same orientation? Then perhaps they were bound into the seafloor, and all swaying in the same current, when the deadly sandstorm hit.

From this kind of analysis, some things have become clear. The Ediacarans were definitely alive—no geological process can make shapes like theirs. They were also much larger than the microscopic creatures of Slimeworld. And though some look like nothing on Earth, others are uncannily like more modern animals—starfish, jellyfish, sponges and sea pens. These resemblances have led many researchers—Jim among them—to believe that at least some of the Ediacarans were direct ancestors of the complex animals we see today. But everyone admits that shape-matching isn't enough. Though the fossils look complicated, they could still be some strange aggregation of simpler creatures. Over the years people have speculated that each Ediacaran could have been one giant single cell, quilted into many fluid-filled compartments, like an air mattress; or perhaps even some exceptionally coordinated colonies of bacteria, banding together into deceptively complex shapes.[6]

As it turns out, they are neither of these. We now know that Ediacarans truly were the first complex, multicellular animals. The proof has emerged only in the past couple of years, and it comes not from the shape of the fossils, but from their *trails*.

IN HIS collection at the Institute of Palaeontology in Moscow, Misha Fedonkin has some of the best Ediacaran fossils in the world. Misha is a dapper man in his early fifties, with a short, tidy moustache and jet-black hair. His English, like his manner, is fluent. He is charming and suave, often animated but never, ever ruffled. After decades of doing science in Russia, he is also infinitely resourceful. That doesn't just apply to fieldwork. Put him in the residential part of an unfamiliar city, and he will immediately find you a delightful little bar, club or restaurant, just around the corner. When Misha was young, he loved hunting and fishing, but now his heart lies elsewhere. He hasn't touched a gun since he found his first Ediacaran fossil, nearly thirty years ago. "Fossil hunting," he says, "captures your soul."

Misha's fossils come from the sea cliffs of Russia's White Sea coast, near the remote northern port of Arkhangel'sk. The train journey from Moscow takes twenty-two hours, crammed in a cabin with everything for the season: army-surplus tents, ropes and climbing gear, food tins, slabs of butter and cheese. From the port there's another ten-hour journey by boat to the campsite, squeezed on a beach between steep, sloping cliffs of clay and the bleak White Sea.

Occasionally a river has cut a canyon through the cliffs on its way to the sea, and Misha usually tries to camp near one of these for the fresh water it provides. Though the White Sea is a branch of the Arctic Ocean, there is no ice on it in the summer. But the weather can still be grim. When it rains, the soft clay of the cliffs turns to pale, glutinous mud that yanks at your boots, and coats everything it touches. Sometimes there is a storm out to sea, and the water rises up on to the beach in a foaming mass, and rips through the camp.

Good weather, on the other hand, brings that other famous Arctic hazard: the flies. At first sight the taiga (Arctic forest) along the rivers looks impenetrable. Then you realize that the stunted trees are in fact widely spaced, and that the gaps between them are dark with dancing clouds of mosquitoes and black flies. On any fine beach day, these blood-hungry beasts will come for you. Like Paul Hoffman in Canada, Misha and his co-workers have gradually become inured to this menace. But one American researcher who joined Misha in the White Sea a few years ago was so badly bitten that his face quickly swelled to twice its normal size. Still, he wasn't particularly troubled. That same year he discovered a new species of Ediacaran that is now named after him. What are a few insect bites, when you can achieve immortality with the neatest of twists—passing your name on not to a descendant, but to an ancestor?[7]

The White Sea cliffs are packed with Ediacaran fossils. Like their Australian cousins, the creatures preserved here lived in a warm, shallow sea, and were suffocated with periodic blankets of sand. Now the crumbly clay of the cliffs is interleaved with layers of sandstone bearing the familiar Ediacaran death masks. Each spring new landslides send sandstone slabs tumbling down on to the beach. Each summer Misha returns to see what spectacular new finds have been loosened by the rains.

And some of Misha's more recent discoveries baffled him. He found four *Dickinsonia,* all exactly the same size, grouped together on a single slab. Puzzlingly, three were raised up proud from the sandstone in positive relief, and only the fourth was the usual indented mould. He also found a slab bearing *Yorgia,* another oval creature, which had internal riblike structures and strange squashed shapes that could have been some kind of organs; there, once again, Misha saw four fossils together, three of them in positive relief, one negative. And there was *Kimberella,* a creature the shape of a teardrop, with a flouncy frill around its edges that looked to Misha like the undulating foot of a slug or snail. At the pointed end of one fossil, Misha found grooves in the rock, as if something had been raking the seafloor just before the sandstorm hit. At the ends of others he noticed long, dark traces, many times longer than the *Kimberella* itself.[8]

Trails. These were all trails. Misha realized that the four *Dickinsonia* fossils had all come from one individual. Three times this creature had rested on the slimy microbial mat that coated the seafloor, and left an imprint of its belly there. The first three death masks stood up from the sandstone slab because the sand had reached out to fill hollows in the seafloor. Only the fourth was indented—a true mould of an Ediacaran's body. And the same thing applied to the *Yorgia*. Misha's sandstone slabs had captured the three previous belly-prints of the organism as well as its own corpse.

When Jim Gehling heard about these finds, he raced off to look at his own collections. Sure enough, he found an Australian *Dickinsonia* doing just the same thing: three belly-prints and one final fossil. And that meant one thing: these creatures could clearly move.

Perhaps, Jim thinks, the *Dickinsonia* and *Yorgia* were using their belly-flops as a way of feeding, since neither had the benefit of teeth. They would lie on the slimy mat covering the seafloor and gradually consume the bacteria there. "If you lay on the lawn long enough, you'd rot the grass underneath you," says Jim. "And that's a source of food." Obviously, then, the creatures would have to move on when the food was exhausted. Misha agrees, and thinks that the *Kimberella* might also have been moving to eat. Those scrape marks could have been places where it used a proboscis to rake in food from the seafloor. And the travelling *Kimberella* left a long trail in its wake, just like a slug or a snail.

The ability to move sends an immediate message to biologists everywhere. These creatures had to be complex. To make trails, you need tissues that behave like muscles; you have to be a cooperative organism made of multiple specialized cells. Quilted air mattresses can't do it, nor can groups of bacteria. *Kimberella* didn't just look like a snail. It *moved* like a snail. The extraordinary traces from the White Sea prove beyond doubt that Ediacarans really were complex, multicellular animals.

So now we know what Ediacarans were. But *when* were they? The fossils at Ediacara and in the White Sea lived around 555 million years ago. That's an improvement on the Cambrian explosion, but it's still some forty million years after the Snowball. The next step would be to find Ediacarans from nearer the end of the ice.

MISTAKEN POINT is a windswept, godforsaken promontory on the southern tip of Newfoundland. It is surrounded by the barrens: blasted, treeless heaths covered with mosses, lichens and tart crown berries. Nobody could love these barren lands, not even their mother. They are dreary and damp, their plants the colour of overcooked spinach and rusty nails; when the wind is not buffeting them or rain beating them down, they are shrouded in fog. The pale, thin caribou wander over them like lost souls.

The seas hereabouts were once rich with fish, but now that cod stocks have crashed and cod fishing is banned, depression has descended on the area like one of its famous fogs. (This is, officially, the foggiest place in the world.) Nobody has lived at Mistaken Point for decades, and only a few people remain in the village of Trepassey, an hour's drive to the west. The locals there are friendly, their accents a fossilized form of Irish, their grammatical constructions archaic. "There you be. You likes that, doesn't you?" they will say as they hand you a plate of salt beef. Or more probably cod, flown in from somewhere, since old dietary habits die harder than most.

Trepassey means "the dead souls" in Basque, and was named by fishermen of the sixteenth century for the many ships wrecked on this craggy coast. These waters hold the remains of thousands of people who were betrayed over the centuries by fog and Arctic ice and high winds, especially at Mistaken Point, which was "mistaken" by many unlucky sailors for the next finger of land along the coast. Seeking the safe harbour at Cape Race, they would turn too soon, and founder. The lighthouse at Cape Race received distress signals from the stricken *Titanic,* and is the closest place on land to the ship's Atlantic grave.

The rocks of Mistaken Point are also grave sites, but for much more ancient creatures. To see these, on one particularly grim day at the beginning of June, I have joined a troupe of soggy, sodden, steaming geologists squelching over the saturated mosses and the black, muddy streams near the cliff top. The rain is relentless. There is no perceptible distinction between sea and sky—both are the colour of granite. Disembowelled sea urchins lie where they were dropped on the rocks and then eviscerated by seagulls or yellow-headed gannets.

The fossil surfaces extend out sideways from the cliffside like a toppled stack of books. We climb down on to one of them, and huddle miserably in the wind, praying for a gap in the clouds. The rock is dull in the flat grey light, and the surface appears blank. Though we are crouched some twenty feet above the sea, occasional waves crash over the edges of the rock layer and people leap out of the way, squealing. Everyone has reverted to geology talk; they are speaking of clasts and volcanic bombs, gravity flows and forearc basins. Someone is talking about Paul Hoffman, though he isn't on the field trip. Bragging about him. "Paul worked for me once. He was my junior assistant in the field." "Did you hear that?" my companion says. "Everyone wants to claim a piece of Paul."

The afternoon is drawing on, but it seems the rain may be easing. Now the wind is a blessing, as it dries off the rock surface. And then suddenly, miraculously, slanting rays of sunlight appear through the clouds. Fossil light! The surface of the rock is suddenly crammed with strange shapes: fronds and spindles and discs and branches. "Look at that!" Jim Gehling whoops. "Someone switched on a projector in the sky!"

Despite the damp rocks, everybody immediately takes off their boots. These fossil forms may have survived thousands of years of pounding by wind and waves, but nobody wants to risk scratching them. As geologists pad over the rock surface in their thick hiking socks, the scene seems oddly biblical. And yes, there's even a "burning bush" fossil, the size of a spread hand, with fronds curling upwards like tongues of flame. Jim is standing to one side, staring raptly at the shapes that have appeared in the rocks. "Just imagine if you could have this as the floor of your house," he says. "Imagine if you could walk on it *every day.*"

The fossils of Mistaken Point are different in many ways from those at Ediacara or the White Sea. These creatures were nowhere near a sandy shore when they died. Instead they rested on the deep, dark floor of an underwater canyon. And they were overwhelmed not by sand, but by ash.

Northwest of here, a range of volcanoes once poked through land that would later become parts of Central America and Brazil.

Now and then, there would be a rumble, a roar, and an explosive eruption that sent dark clouds of ash flying through the air and into the ocean. The Ediacarans had no ears to hear any warning sounds. A thick, swirling cloud would simply have appeared from nowhere and filled their world, blanketing the canyon floor and everything that lived there.

The ancient Roman town of Pompeii was smothered by just such an ash cloud when Vesuvius erupted in 79 A.D. Fleeing, crouching or writhing, the bodies of Pompeii's residents were preserved in death by the same ash that ended their lives. First this ash smothered them, then it hardened around their rotting corpses. Centuries later, the bodies themselves had disappeared. But archaeologists injected plaster into the voids left behind, and re-created the shapes of the dying residents in extraordinary, sometimes horrible, detail.

In just the same way, volcanic ash hardened over the rotting bodies of the Ediacarans. Then, like the plaster at Pompeii, mud from the former seafloor forced its way up into the hollows they left behind, making faithful images that turned slowly into rock.

Preservation by ash is unusual. Most Ediacaran fossils were smothered by sand, and their images are preserved only as indented death masks in an overhead layer of sandstone. But in Newfoundland the image comes from the underlying mud. The ash layers have mostly weathered away, and solid casts of the Ediacarans stick up from the rock surface. Mudstone on mudstone, they are still only visible when the sun is low enough to outline them with their own shadows. But then they are revealed exactly as they once lay—Newfoundland is the only place in the world where you can walk around on the Ediacaran seafloor.

This rare preservation method has another benefit. Each layer of ash provides the Ediacarans it killed with a handy age marker. Volcanic ash is a great way to date rocks, because it contains traces of radioactive elements; uranium, for instance, which decays into lead at a precise, well-defined rate. When a volcano explodes and its ash forms, this uranium clock starts ticking. With each tick, the rock loses uranium and gains lead, and the ratio of these two elements gradually changes over time. In an ash layer, geologists can measure the ratio of uranium to lead today, and can tell how much time has passed since the ash formed.

Here at Mistaken Point, the ash that smothered these jellied fronds and spindles gives an age of 565 million years.[9] The creatures here are much older than the ones in Australia or the White Sea. They lived just 25 million years or so after the Snowball.

They're not even the oldest fossils in Newfoundland. The next day we find ourselves on another part of the coast, this time close to a beach where each receding wave sucks the pebbles backwards with a sound like a crackling flame. The rock layers here are giant black slabs, tilted sideways like collapsed dominoes. We head for one that has some kind of large discs etched into it. "Pizza to go, anyone?" says Guy Narbonne, the Canadian researcher who is leading the trip. He's right. These fossils look exactly like pepperoni pizzas. Or at least they are the perfect size and shape; but the pepperoni pieces are mud-coloured and the region around them is apple green, like slightly mouldy dough. A careful hike over the slippery rocks takes us to a red surface, which bears the faintest trace of what looks like a long-stemmed cocktail glass. Near it are two thin fronds, several feet long. At least these don't remind anyone of food; they look more like the prints from a bicycle tyre.[10]

Where the rock layer disappears into the ground, we can see, edge-on, the ash layer that preserved these faint fossils. I'd half expected it to be dark and crumbly, but instead it's as solid as concrete, and the same pale green colour as the pizza "dough". This ash has been dated. The result bears the status of a "rumourchron"—geologist slang for a date that has been measured but is not officially out in the world. It comes from the lab of Sam Bowring at MIT, one of the most reliable geochronologists in the world. Sam has told the date to many people. He's presented it at conferences.[11] But he hasn't quite published it yet. It's 575 million years. If his date holds, these fossils are almost as old as the final days of the Snowball.

STEP BY step, the date for the invention of complexity is approaching the end of the Snowball. For many people this is beginning to look like more than a coincidence. But one serious problem still remains. A few researchers have turned up what they think could be signs of complexity *before* the Snowball. If multicellular life really did emerge before there were even glimmers of global ice, that could scotch the whole biological part of the Snowball theory.

Some of the evidence is still highly controversial. An outcrop in northern Canada bears perhaps a thousand discs, somewhere between the size of a dime and a quarter, impressed into the bottom of sandstone rocks. The rocks date from around 100 million years before the end of the Snowballs, and their discoverer, Guy Narbonne, insists that they are complex creatures.[12] But they left no trails, and have virtually no structure. Other biologists say they could well have been simple blobs of jelly or colonies of bacteria.

And etched on billion-year-old rocks from India are pencil-thin branching tubes, which their discoverer—a most respected researcher named Dolf Seilacher—believes were made by some kind of early worm.[13] But most of his colleagues sigh and point out that there's no sign of the creatures themselves among the "trails", which makes his argument much harder to swallow. Other researchers have just reported the discovery of blobby grooves, like worm casts, in 1.2-billion-year-old sandstones from southwestern Australia.[14] But once again, there's no sign of any animal, nor any clear evidence that complex creatures really created these "trails", and few biologists think they pose problems for the Snowball theory.

More troubling, though, are the algae. Algae are marine plants that live throughout the modern ocean. Some are small, hairy blobs floating through the water or clinging to rocks. Others are huge. Kelp is a form of algae, and the kelp forests off the coast of California contain plants that are hundreds of feet tall. Algae certainly existed before the Snowball. There was nothing like kelp; the biggest creatures were just a fraction of an inch across. But they were almost certainly multicellular.

For instance, Nick Butterfield, a Canadian biologist now at Cambridge University, has found fabulously preserved red algae in a lump of chert that he collected from Somerset Island in the Canadian Arctic. The rock is 1,200 million years old, and the fossils it contains are tiny, hairy things, scarcely visible to the naked eye. But when Nick put his samples under a microscope, he realized that the fossil images were dead ringers for a modern red alga called *Bangia,* which you can scrape off rocks on many seashores today. He saw the classic rows of disc-shaped cells that make up the *Bangia*'s filaments, and the wedge-shaped cells that adult *Bangia* possess, having divided their discs into eight, twelve and sixteen pieces. He also saw separate cells that were orientated vertically, and appeared to be making up a "holdfast", a kind of anchor that could bind the alga into the seafloor and enable it to grow upwards rather than merely sideways like the primitive flat mats of Slimeworld.[15]

Then there's fancy filamentous algae from Spitzbergen, which look much like green algae does today. And a strange beast named *Valkyria,* with appendages that look almost like legs (but aren't). And a Siberian fossil found by Andy Knoll, a colleague of Paul Hoffman's at Harvard, which looks just like a modern green alga called *Voucharia.* Many of these are not just collections of cells. They really look as if they've already learned to specialize.

These finds may seem to topple the biological part of the Snowball argument, but they leave open one big mystery. Algae apparently learned to be multicellular by 1.2 billion years ago. If they then passed on the secret to the rest of the world, why did it take another 600 million years before animals did the same? If this was the crucial step that changed the world for ever, why did the rest of the planet stay mired in simple slime for so long afterwards? *Nobody* believes that one evolutionary event can trigger another occurring hundreds of millions of years later. Even Nick Butterfield says so. "There's still this huge delay before things got rolling," he acknowledges, rather sadly. "Biology moves faster than that."

There are, then, two remaining possibilities. Either algae invented complexity separately, and kept the secret to themselves—which would pose no particular problem for the Snowball idea—or there were plenty of complex animals around before the Snowballs, but they left no trace in the rocks. That would definitely be a problem, since the Snowball couldn't trigger something that already existed long before. But how would you test this without fossils? There might just be a way. The evidence would come not in the form of fossils, but from applications of a more oblique approach known as a "molecular clock".

The genetic material—the molecule called DNA—inside every living cell contains information about its ancestry. In principle, with a sample of my DNA and some of yours, we could work out how closely you and I are related. Although we are both humans, your DNA differs slightly from mine. That's why our faces are perhaps different shapes, or our eyes a different colour. These changes in DNA have happened over many generations, as genetic material passed from parent to child, sometimes with small mistakes introduced, sometimes just through the natural mutations that appear over time.

When my DNA was last identical to yours, it resided in the cells of the person who was our last common ancestor, our mutual great-great-ever-so-great-grandparent. So if we wanted to work out when this ancestor lived, we wouldn't necessarily have to consult a genealogist. Instead we could simply measure the differences between my DNA and yours, and estimate how quickly the DNA clock ticks.

In practice, DNA changes aren't fast enough to help with recent family trees, though researchers have used this technique to show that we are all descended from one modern human "Eve," who lived a little more than 200,000 years ago. Molecular clocks can also identify the timing of more-distant ancestors—between a human and an orangutan, say, or a human and a fruit fly. Each tick of the clock, each slight change in the exact composition of two creatures' DNA, takes them further away from their mutual ancestor. If you measure how much their DNA has changed, and you know how fast the clock was ticking, you can track evolution backwards even without the help of fossils.

Many different research groups had already used this approach to try to find out when complex animals first emerged. They examined the DNA in different animal species alive today, and then worked backwards to try to date the appearance of their last common ancestor. The first of these studies, back in 1982, said this unique animal ancestor lived around 900 million years ago. Though others thought it might have been a little younger, the most recent studies have pushed the date much further back. In the past couple of years, several different molecular clocks have suggested that the animal ancestor lived some 1.2 billion years ago or more, long before the Snowballs had even begun.[16]

Kevin Peterson is a fervent young biologist from Dartmouth College in New Hampshire. He doesn't like the Snowball idea. He doesn't really like *any* big idea. What he cares about, he says, are hypotheses. Unlike ideas, hypotheses are *testable*. Kevin is very big on testability. If you told him you thought tomorrow was Saturday, he'd probably ask if your hypothesis was testable. Unless he can test something, he doesn't want to know.

To some extent, this is true of all scientists. Speculating is fun, but if you can't decide whether one speculation fits reality better than another, then why bother? "However beautiful we may find the constructions of our imagination," wrote the physicist Lee Smolin, "if they are meant to be representations of the natural world, we must take those constructions humbly to nature and seek its consent."[17]

There's nothing particularly humble about Kevin. He's a

young Turk, supremely confident that his own findings will shoot down those of his scientific seniors. But he certainly believes in consulting nature. And as soon as he saw Paul and Dan's papers about the Snowball, he was revolted. Yes, yes, lovely idea, but where's the *proof*? Kevin wasn't concerned with the geological side of the argument. That's not his speciality. But when he saw the part about biology, he was infuriated. Paul and Dan had drawn a diagram showing the first complex animals appearing after the Snowball. "I thought it was one of the silliest figures I'd ever seen in my life," Kevin says. He knew all about the molecular clocks. And every single one of them said that complex animals formed hundreds of millions of years before the Snowballs even began.

The problem was that these previous attempts produced a disturbingly wide spread of dates. Kevin decided that was because they hadn't been done correctly. There were flaws, he felt, in each of them. So he resolved to do a new, improved study. He would design the best-ever molecular clock. He would use it to pin down the timing of the animal ancestor once and for all, and—he presumed—to demonstrate that the biological part of Paul's idea was hopelessly wrong.

Kevin chose the creatures for his molecular clock carefully. He decided on echinoderms—the family that contains urchins and sea stars. For one thing, there was a complete, well-dated set of fossils for the ancestors of these creatures, so he could check more carefully than previous studies how well his clock was doing as he went backwards in time. For another, these creatures had similar body size, metabolic rate, and amount of time between one generation and the next. Kevin felt that previous studies had erred by picking creatures that were too different in all these respects. The more alike the creatures that he started with, Kevin realized, the more accurate the clock tracing their mutual ancestry should be.

So he set about uncovering the genetic sequence behind seven different creatures. He calculated how fast the DNA must have changed. He worked steadily backwards, checking his dates as he went. Whenever he had well-dated evidence from fossils for the timing of a particular ancestor, he checked whether the clock agreed. Each time, the clock looked good. Encouraged, Kevin projected his clock further back in time. Now there were no fossils to check against. Now he was getting closer and closer to the animal ancestor. And then, finally, he had his answer. The last common ancestor of all complex animals lived . . . somewhere around . . . 700 million years ago.[18]

Kevin was stunned. "I couldn't refute that diagram of Paul and Dan's," he told me several months later, still sounding dazed. Why not? After all, his clock didn't throw up that magic date of 590 million years—which marks both the ending of the ice and the beginning of the first complex fossils. But neither did it prove that animals had existed more than a billion years ago, as Kevin had expected. Instead, its date agreed almost exactly with the end of the very first Snowball.

Remember that Paul was dealing with a series of events, not just one. And the first Snowball ended around 700 million years ago, exactly the date Kevin's clock produces for the animal ancestor. Perhaps complex animals were triggered by the first Snowball, and then survived through the remaining episodes of ice. If each subsequent Snowball wiped out all but a few of those new animals, that might explain why widespread fossils didn't appear until the ice finally receded.

It's also possible that even Kevin's careful clock overestimated the age of complex animals. Genetic studies like these assume that their clocks have always ticked at the same rate. But some biologists think that genetic changes happened more quickly in the past, and that all the clocks give older times than they should. Kevin believes he solved many of the problems with the earlier clocks, but he says himself that he may not have solved them all.

THERE WILL be many more attempts to find traces of the earliest animals. Researchers are collecting genetic material, designing newer, better molecular clocks, and scouring the world's rocks for trails blazed by ancient life. But the more biologists try to pin down the timing of widespread complexity, the more their results seem to point towards the Snowball.

Was this just a coincidence? "Most people I know think they're connected," says Jim Gehling. And Kevin has changed his mind about the timing at least, though he still wants a testable hypothesis to explain the connection. Nick Butterfield, the algae man, says the same. He is annoyed about Paul's airy biological assumptions. He says that Paul's paper was "cringingly awful in its biology". He says it's up to Paul to explain exactly how ice could have triggered life's industrial revolution.

But forget Paul for a moment. What does Nick think about the biological evidence, the new fossils and the molecular clocks and all the other developments that are steadily moving the date of burgeoning complexity closer and closer to the ice? Nick pauses. Then says this. "I think it's *fascinating.*"

Biologists aren't so very different from geologists, under the skin. Some of them have jumped eagerly on to Paul's Snowball bandwagon, and some have declared furiously that it must be stopped. And some are still waiting to see what will happen.

Though the world of ancient fossils seems pretty well explored, it's still possible that someone, somewhere, will find a vast stash of complex animals from long before the Snowball. But in the absence of this, the evidence for some connection between ice and new life is looking more persuasive by the day.

So biologists are beginning to think of ways that the Snowball might have triggered complexity. Everyone agrees the capacity to be complex must already have existed in the creatures' genes, but nobody knows for sure what spurred those dormant genes into action. The theories are not yet fully formed; they're speculations in corridors rather than neat, tidy theories. But there are several intriguing ways in which ice, that most inimical of substances, might ultimately have given rise to this new life.

The Snowball itself could have encouraged life to diversify and experiment. New species often arise when a single population of creatures is separated from its fellows in an isolated refuge, for something upwards of a million years. Or perhaps the opportunity for complexity arose after ice wiped large areas of the planet clean of life. All living things need certain resources to survive—food, water and shelter—and as long as enveloping mats of slime were hogging all the resources, there would be no space left to innovate. Removing the extant occupants of Earth's ecological niches might have made room for life to experiment. We already know that worldwide extinctions make space for new species to emerge. When a meteorite wiped out the dinosaurs, for instance, the previously tiny mammals suddenly had free licence to grow, change shape, and consume the resources once reserved for the likes of *Diplodocus* and *T. rex.* Though there's no direct sign that the ice made any of the slime-creatures *extinct,* it may have killed off enough of each species to create the breathing space that evolution needed.

Another suggestion has come mainly from Jim Gehling. He wonders whether complex life was a response to the sheer changeability of life in the Snowball's aftermath. First the world endured its longest and most severe ice age, and then came a violent hothouse lashed with acid rain. With conditions changing as drastically as this, life had a natural incentive to spawn creatures that could protect themselves from external buffeting. Single-celled slime balls are at the mercy of current and weather, but large, multicelled animals have much more control. They can dig into the ground and hold tight in fierce currents. They can control their internal temperature, store food more effectively against lean times, and grow covers to protect themselves.

But the most popular idea for a trigger point involves oxygen. Large creatures need efficient ways of mobilizing their food into energy, and oxygen is one of the best. When we breathe, the oxygen we inhale is used to "burn" food, like burning petrol in a car engine, and that's what generates the energy that supports our vigorous lifestyles. Oxygen is also necessary to make collagen—the connecting tissue that binds muscles to bones and helps keep cells together, and that is found somewhere or other in every complex animal.

There are some signs in the rocks that atmospheric oxygen was increasing around the time of the ice. Perhaps whatever triggered the Snowball also created this excess of oxygen. Or perhaps there was a sudden pulse of oxygen immediately after the freeze ended. For millions of years, life would have been restricted to a few small refuges, and unused nutrients would have built the ocean up into a tasty chemical soup. As soon as the ice was over, the few remaining creatures would have seized on these nutrients and blossomed. The white planet would have become green with massive colonies of bacteria and algae stretching over the surface of the ocean. And those same colonies would have soaked up sunlight, made food, and belched out oxygen as a waste product of their endeavours. That sudden pulse of oxygen may have been exactly what complex life was waiting for.

Biologists are now trying to figure out how to test these ideas. But there's something they all agree on. Whatever creative role the Snowball may have played in shaping a new world order, it would also have been devastating for many of the life-forms it first encountered. And this raises a disturbing question: could another Snowball happen today? If the ice returned to haunt us, the consequences would be horrific. Earth has come a long way since the simple days of Slimeworld, and life is now a complex web of interdependent creatures. If another Snowball engulfed the Earth, many—perhaps most—of these creatures would perish.

TEN

EVER AGAIN

To figure out whether the ice will ever return, we first need to know why it appeared in the first place. What was so special about the Snowball times? Though the clues are scant, some evidence has emerged from another, even older part of Earth's geological history. Joe Kirschvink, the sparky, inventive Caltech professor who set much of the early Snowball rolling, has discovered that Paul's Snowball period wasn't the only one.

SOUTH AFRICA, SEPTEMBER 2000

You'd EXPECT the Kalahari Desert to be dry and hot, and so it usually is. Even the place names around here evoke its baking, insufferable summers—Hotazel, for instance, a remote mining outpost an hour or two north of here. (The land surveyor who proposed the name back in 1917 had to use this phonetic spelling because the authorities objected to his original suggestion: "Hot as Hell".)

But now, at the tail end of a southern winter, the desert is both cold and very wet. The rain began yesterday evening with a roar that eventually settled down into a night-long drumroll on the tin roof of our tiny motel. It has turned the dirt road into a skating rink of rich, red mud. I'm here with Joe Kirschvink, who has brought a phalanx of students to tour the geological sights of South Africa. As our five vehicles lurch in convoy through the puddles, fountains of red water fly into the air and separate out into thick, cartoonlike drops.

We turn, thankfully, on to a paved road again, and the rain begins to ease. The landscape in the southern Kalahari is just like Paul Hoffman's Namibian field sites: open and almost featureless, scattered with camel thorns and golden grasses and those towering red termite mounds, as tall as the trees. There's a striped gemsbok, sheltering among the thornbushes. And there, hanging from a telegraph pole, is a familiar weaverbird's nest, an impressive sack of entwined grasses nearly five feet long and almost as wide. The road begins to gain altitude, and the temperature drops further. A dense white mist descends around us, hovering off-road just beyond the barbed-wire fence. At the designated road cut, we climb out of the vehicles and wince as flecks of ice fly by in the freezing fog. But Joe is cheerful. "What did you expect?" he says with a grin. "This is Snowball country."

He's right. All around us, among thornbushes and orange clumps of grass, are the now-familiar signs of ancient ice. The background flame-coloured rock is studded with pebbles and stones that were once bulldozed along the ground by glaciers, and then tipped offshore into a shallow, ice-covered sea. Not all of these have remained embedded. The sides of the road are scattered with loose pebbles and we spread out, red-cheeked in the wind, seeking stones with tell-tale glacial scrapes. Here's one, a small rounded lump, indented with parallel grooves where it was scoured along the ground. These deposits bear all the hallmarks of a Snowball. Not just the jumbled, scratched ice rocks that we see here; back at Hotazel there are also thick layers of iron that rusted out of the Snowball ocean, and above them, the classic cap carbonates that brought Dan and Paul their eureka moment.

But this Snowball differs from Paul's in one crucial respect: it happened nearly two billion years earlier. These are the remnants from a Snowball that gripped the Earth not 600 million years ago, but a full 2.4 billion years ago. Rocks this old tend to have suffered greatly from the tectonic twisting and weathering that billions of years inevitably bring; intact outcrops are rare from the Earth's early history, and even the few that have survived are tremendously hard to interpret. So although geologists have known about these truly ancient signs of ice for decades, nobody has paid them much attention. Unlike the ones from Paul's more recent Snowball, these ice rocks don't appear on every continent, and not many bear the clear tropical hallmarks that gave Snowballers their first clues. Nobody would have guessed that these few deposits marked another, earlier global ice age. Until, that is, Joe came along and proved it, using yet another of his magic magnetic measurements.

Back in the warm trucks, we head for the site of these measurements: a road cut that's about an hour's drive away. There the route has sliced through slate-grey remnants of ancient volcanic eruptions. The deposits are immense, a thousand feet thick, and they once covered much of the Kalahari. They are "flood basalts", so called because the lava gushed from the ground in torrents for perhaps a million years, and created a smouldering new surface. These volcanics, Joe tells us, lie right in the middle of the Snowball rocks; they emerged from the ground into a world that was already gripped by ice. "It must," Joe says, "have been wild."

Above our heads was once a shallow, frozen sea, its shoreline just a little way over to the east. Beneath our feet, lava was issuing from cracks in the seafloor, heating the water, melting the overhead ice, and filling the air with hissing clouds of steam. The sea's icy surface was pocked with hot pools, which were green with grateful clumps of slime. As the lava flooded out into the cold seawater, it cooled immediately into gloops of rock like toothpaste writhing from a tube. These rocks are classic signs of an underwater eruption. Geologists call them pillow lavas, but they look bulbous and blubbery—more like elephant seals lolling on a beach. Here, they are collectively named the Ongeluk Formation; *ongeluk* means "misfortune" in Afrikaans, but they brought good luck for Joe. Using these rocks, he discovered that when this part of the Kalahari was coated in ice and fire, it rested within a few degrees of the equator.

Remember that many rocks carry a magnetic birth certificate. When they are young and soft, any magnetic minerals they contain will line up like tiny compass needles along the Earth's local magnetic field: vertical near the poles, and horizontal near the equator. And when the minerals harden into rock, this pattern is frozen in. Wherever these rocks wander, they take their birth field along with them.

So, simply measuring the magnetic field in the rocks should tell you where in the world they formed. There is, though, a catch. Remember, too, that subsequent events—heating, mangling or importing of new magnetic minerals—might have overwritten the original pattern, issuing the rocks with a fake birth certificate. To find out where the rocks were born, Joe didn't just have to measure their magnetic field; he also needed to prove that the field was original.

Joe realized that the pillow lavas could help him do this. The field test he planned was a little like the fold test he used to check whether the Australian rocks had a genuine magnetic memory. But in this case he wasn't seeking a fold in the rocks. Instead he wanted to find shattered volcanic shards. As red-hot molten rock spills into cold water to make pillow lavas, the outer surface immediately freezes in shock and becomes a brown, glassy coating called a "chill margin". The inner parts of the rock take a little longer to catch up. When they finally do cool, they shrink, and this process often shatters the brittle outer skin.

The chill margin takes on the local magnetic field the moment it cools. If it then shatters and the shards point in random directions, their magnetic arrows should be just as random. But if the magnetic field of these rocks has been overwritten after they formed, cooled and shattered, the fields in pillows, shards and everything else will all point neatly in one overall direction. Joe knew that if he measured the fields of the chill fragments, he would be able to tell if the magnetic birth certificate was original.

Here at the Ongeluk, we can still see the small cylindrical holes where Joe took his samples. Joe calls them "palaeomaggot holes", but he made them himself with a rock drill almost a decade ago. He eventually analyzed the results years later, at the prompting of a particularly insistent graduate student. And he then discovered that the volcanics, and the ice rocks that bracketed them, contained a field that was almost flat. What's more, the shards of shattered chill margin had fields that pointed in every direction. This meant the measurement was genuine, and that 2.4 billion years ago, ice lay within a few degrees of the equator. These ice rocks, in other words, are just like Paul's. They are the remnants of another, earlier Snowball.[1]

So NOW we know that Snowballs have happened twice. At least one occurred a little over 2 billion years ago, and then a series of perhaps four engulfed the Earth between 750 and 590 million years ago. There have apparently been no others. What, then, did these Snowball periods have in common, and what made them different from every other time period in Earth's long history? Was there anything unusual about them that could have triggered the ice onslaught?

Perhaps. There are intriguing magnetic hints that both of these time periods had a peculiar continental alignment. As the world's tectonic plates drift over its surface, the continents sometimes bunch and sometimes scatter. When they spread out, they can end up anywhere. But on a few rare occasions, they can find themselves in a band around the Earth's equator. And this might be exactly what happened during the Snowball periods.

Though magnetic measurements are difficult, and many of the sites have had their magnetic memories rewritten in the intervening time, decent data exist from about half the continents that were around during the later Snowballs. And every one of these lay near the equator. So, too, did the half-dozen sets of ice rocks that have now been measured. For the earlier Snowball, the task is harder and the measurements are fewer. But still, all of them point to low-latitude continents.

If the continents truly were arranged around the equator during these two Snowball periods, that could be just what the ice needed. One reason is that the tropics soak up most of the heat that arrives on Earth from the sun. Because land is more reflective than ocean, putting all available land in the tropics could reflect more of the incoming sunlight, and help the planet to cool. Joe Kirschvink suggested this in his short paper back in 1992.

Dan Schrag, the ideas man, has come up with another reason why equatorial continents could be the key. When continents spread out to the far north and south, he says, they act as an important brake on overenthusiastic polar ice caps.

Ice naturally wants to spread: white ice reflects sunlight, which causes cooling, which breeds more ice—and if this were left unchecked, Earth would spend its entire life as a Snowball. Fortunately for us, high-latitude continents stop that from happening by helping to warm the Earth back up again whenever polar ice becomes rampant.

Normally, rocks do the opposite. They help prevent the Earth from overheating by soaking up the greenhouse gases like carbon dioxide that are pumped out by volcanoes. But if polar ice starts to spread, any high-latitude continents will switch loyalties. Because their rocks become covered with ice, they can no longer soak up carbon dioxide. Instead it stays in the atmosphere to do its greenhouse thing, warming up the Earth and melting the excess ice. So if ever the polar caps start to grow, high-latitude continents will force them to shrink again.

Now imagine what would happen if all the continents were arranged in a band around the Earth's equator. In that case the polar ice caps could spread with impunity. There would be no high-latitude continents to cover, and hence nothing to stop the ice going all the way. By the time ice reached the equatorial continents, it would be too late to prevent a Snowball.[2]

This idea also neatly explains why, at least in Paul's period, there was a series of Snowballs rather than just one. Between 750 and 590 million years ago, the continents could simply have stayed near the equator. A Snowball would begin when some trivial cooling trigger set the ice moving. With no high-latitude continents to stop it, the ice would continue until the Earth was encased. Over the next 10 million years or so, carbon dioxide gas pouring out of volcanoes would build the atmosphere into a furnace, until it became so hot that the ice melted back. Gradually, then, the carbon dioxide levels in the atmosphere would drop, until the whole process started again. As long as the continents stayed near the equator, another cooling trigger would set another Snowball rolling. And another. And another. Until, eventually, the continents moved on and the world was spared.

There aren't enough outcrops from Joe's earlier Snowball to know whether it was a series of events or just one. But researchers believe that the sun was much feebler then, and that an individual Snowball would have lasted much longer. Perhaps Joe's single early Snowball lasted so long that the continents had begun to move away from the equator again by the time the ice receded.

So equatorial continents could provide the rare but reasonable recipe for a Snowball. If this explanation is right, that's encouraging news for our own future. Right now we have plenty of continents at high latitudes. Most of the world's landmasses are way up in the north—think of Canada, Europe and Russia. Presumably, these far northern lands are protecting us all from the ice. Well, not necessarily. It turns out that in spite of this reassuring continental arrangement, the Earth may even now be preparing for another descent into ice.

DAVE EVANS used to be a graduate student of Joe's. He's the one who prodded Joe into measuring the South African samples, and proving that the older ice rocks had been close to the equator. (Dave found the samples collecting dust, and resurrected them.) Now, in his early thirties, he's a professor at Yale University. He is thin and gangly, with thick, wavy hair and a pleasant smile, and looks younger than most of his students. Though he is organized and careful, from working in Joe's lab he also has this legacy: the capacity to consider crazy ideas that might just be true.

While he was at Caltech, Dave didn't just work with Joe on the ancient Snowball. He also investigated another of Joe's "nutty" ideas. As the Earth's tectonic plates creep over its surface, they usually travel at a sluggish few inches a year—the same speed that your fingernails grow. But Joe and Dave believe that at certain times in the past, the continents let rip, travelling at what for them was the breakneck speed of several feet a year. They did so, according to this theory, because they had an inexorable urge to reach the equator.

Spinning objects always prefer to have most of their weight around their middles. Think of a child's spinning top: tall, thin ones are much easier to knock over than short, fat ones, because they're more unstable. If the instability is too much to handle, the object will try to readjust. Suppose you dropped a large lump of clay on to the top of a spinning basketball. If the lump was heavy enough, the basketball would tip over until the excess weight was spinning around its waist, and the system was safely back in balance.

Joe and Dave believe the same thing applies to our spinning planet. They think that if the shifting continental lumps on its surface throw it off balance, the Earth will try to move them equatorwards. This doesn't happen all the time, they say. There has to be a big enough imbalance before the Earth will notice. But occasionally the random jitterbug of the continents brings them crashing together into one massive "supercontinent".

Even that's still not quite enough. All the continents in the world don't weigh much compared to the Earth's massive innards: its thick mantle of plastic flowing rock and its hefty iron core. But Dave believes that the supercontinent would act as an insulating cap over the mantle that lies beneath. Gradually the mantle would heat up, and a great plume of rock would rise up like lava beneath the supercontinent, lifting it up like a giant pustule. Now, with continents and mantle together, the lump would tip the balance and the Earth would respond. Both supercontinent and underlying mantle would go flying off to the equator, until the world became stable again.[3]

This idea is every bit as controversial as the Snowball, and Dave—one of the few people in the world to work on both—has now thought of an ingenious potential connection between the two theories. What if slipping supercontinents make Snowballs? First the continents would collect together into one enormous mass; then this mass would skid to the equator; then the hot plume of rock that still lay beneath the supercontinent would blast it apart in a frenzy of volcanic activity that left fragments scattered around the equator and tropics. And while this was going on, the polar caps could proceed, unchecked, to cover the Earth.

At least some of the available evidence fits this idea. A supercontinent that geologists have named Rodinia finally broke up around 750 million years ago, exactly when Paul's Snowball episodes began. Nobody knows whether there was a supercontinent before the earlier Snowball, the one whose remnants Dave and Joe measured in South Africa. But those same bulbous pillow lavas that provided their samples might also contain clues about the state of the continents then. Those massive volcanic floods didn't cover only South Africa; they also poured out on to many other parts of the world. And that's exactly what you'd expect if a supercontinent was breaking up, and huge amounts of lava were spilling through the cracks.

Were there any supercontinents that didn't produce Snowballs? Well, one called Pangaea existed around 225 million years ago, without generating any notable ice. But Dave points out that Pangaea broke up again relatively quickly. He suspects that it simply wasn't around long enough for that crucial plume of hot rock to form underneath, and unbalance the Earth.

Now we're at the outer reaches of the Snowball idea, with speculation heaped on speculation. But this idea of Dave's is intriguing. And if he's right, the corollary is also chilling. You see, the Earth is making another supercontinent right now.

SIXTY MILLION years ago, not long after an asteroid slammed into the Earth to end the rule of the dinosaurs, India began to sense the presence of Asia. The bulk of what is now the Indian subcontinent had been drifting, footloose, ever since it broke away from Antarctica during the shattering of Pangaea. Now it was moving steadily northwards at the rate of a few inches per year, and Asia was in its way. There was only one possible outcome: a continental pile-up. When ocean basins collide, one or other of the crusts tends to be forced downwards, back into the Earth's interior. But continents are not nearly dense enough to sink. When two continents crash, the only way is up.[4]

So India crashed into Asia, and the land began to rise. First the crust of Asia squeezed around the sides of the thrusting arriviste. Then, as India wedged itself like a chisel further beneath Asia, the surface crust crumpled and folded into a range of mountains more than two thousand miles long. These were the beginnings of the Himalayas. And the land around the mountains was forced up into a vast plateau, the "roof of the world", whose average height is greater than the highest mountain in America. India is still pushing. The Himalayas grow by nearly half an inch a year, and Everest and its kin would be even taller if their fresh young rocks weren't eroding away as they rose.

Meanwhile, partway round the world, Africa was aggressively reacquainting itself with its old Pangaean neighbour, Europe. The first part to hit was a peninsula, sticking out from the northern part of the African plate and bearing what is now Italy and Greece and the countries of former Yugoslavia. This collision threw up the beginning of the Alps. Spain crammed into France, and henceforth there were Pyrenees. And though Africa and Eurasia may seem as if they are only joined at their Arabian hip, the Mediterranean is slowly closing. When Africa itself collides with the European continent, a mighty new range of mountains will be born.

Arabia is now shoving into Iran. Europe and Asia have never been parted since Pangaea, and Australia is heading northwards to join in. In a few tens of millions of years, Australia's left shoulder will probably catch on the southernmost islands of Southeast Asia. It will twist and jerk upwards, to slam into Borneo and the southern parts of China.

Predicting the future of continental movements is an inexact science. But supercontinents come and go over time, and most of the world's landmasses are already crammed into this one gigantic block. Only the Americas and Antarctica remain aloof. As the Atlantic Ocean widens, America is moving steadily further from Europe, and the most dramatic drift within the North American continent is the one taking Los Angeles and Baja northwards. (In about 10 million years, L.A. will pass San Francisco, and by 60 million years from now, it will be heading down a trench into the Earth's interior, just south of Alaska.) But some researchers predict that the Americas, too, will be reunited with the rest of the world's landmasses. According to one attempt at constructing the future, over the next few hundred million years the Atlantic will begin to close again, bringing North and South America back into the fold, and Antarctica will head north to join India.[5]

If so, in 250 million years, the new Pangaea could form. Then it would need to survive intact for another hundred million years or so, while a plume of hot mantle built up beneath. It would shift to the equator in a geological eye blink, just a million years or so, to right the balance of the spinning world; it would break up, and scatter its pieces around the equator and tropics. And then the ice would return.

Gradually the frozen polar oceans begin to reach out with tentative feelers of ice. Finding nothing to stop them, they continue their spread. The whiteness advances like a disease that gradually covers the planet's blue surface. The oceans turn greasy, first, with smashed ice crystals. Then the pancakes of ice are back, and the frost flowers, and the transparent young coating of sea ice that bends with the swell. The ice thickens, and spreads, and thickens some more. By the time it reaches the tropics, it's unstoppable, and in just a few centuries it goes all the way. Global temperatures plummet; rain stops; clouds no longer form. Wisps of ice ripped from the frozen ocean are spread by the wind until they begin to build up on the world's highest mountains. Slowly, steadily, the ice forms glaciers that spill down on to the lowlands. And then the white-out is complete.

WHAT WILL our descendants do? Perhaps they will be so unimaginably advanced that they'll be able to prevent a Snowball. They might be routinely tapping additional energy from the sun, or stopping continents in their tracks. But the Earth is a powerful and stubborn force. She limits our resources, and her geological will is extremely hard to check.

If distant descendants of the human lineage can't stop the Snowball, can they weather it? That's hard to imagine, too. Getting a few simple marine creatures through the ice is one thing, but the complex creatures that inhabit our planet today would be another matter. Antarctica is the most hostile place on Earth. Unless you take your own life-support system of food and fuel and shelter there with you, you will die. And in a Snowball, Antarctica takes over the world. For any truly complex creatures, the result would surely be disastrous. Norse mythology has a word for it. After the catastrophe of Fimbulwinter comes Ragnarok, the end of the world.

But a new Snowball wouldn't be the end for *all* life on Earth, any more than the previous ones were. The destructive power of the last Snowball was followed by an extraordinary new beginning. Who knows what direction a post-Snowball Earth might encourage its living things to take?

Our planet is, after all, a master of invention. Through geological time, Earth has constantly sought out new forms and taken on remarkable new identities. Plumes of hot rock ascending from the interior drive continual reshaping of the continental surface. A mountain range rises; another falls. Oceans open here and close there. Earthquakes and eruptions and tidal waves that seem so catastrophic to us are all just part of Earth's irresistible transforming urge. Even the flimsy atmosphere plays its part in adapting, then reinforcing, our planet's shifting moods. Change doesn't alarm the Earth; it is a fundamental part of its nature. We humans, and the other creatures that share our geological slice of life, are the fragile ones.

EPILOGUE

> He walks and walks and walks until
> he's reached the summit of the hill.
> There he rolls a ball of snow
> and aims it at his friends below,
> But then he slips, so now poor Paul
> becomes a part of his own snowball![1]

When Paul Hoffman saw the children's cartoon book containing this poem, he couldn't believe it. Now he has both cartoon and poem proudly taped to his office door. The Paul of the story ends up careering down a hill, trapped inside a snowball, and wearing a look of horror along with his muffler and galoshes. But Paul Hoffman is delighted to be bound up in his Snowball theory. Remember how in 1991, before he had even gone to Namibia, his old university had asked what he would like to be remembered for. And how he'd promptly replied, "Something I haven't done yet." If you present him with the same question today, he hesitates. "I suppose I should say the same thing—something I haven't done yet," he says at last. "But the Snowball's going to be pretty hard to top."

Paul has won the medals he sought. He's particularly proud of the Alfred Wegener medal, awarded by the European Union of Geosciences for research that successfully brings together many different fields "in the spirit of Wegener", that other charismatic, vilified champion of a theory that rocked the world of science. Paul received the medal in April 2001, and gave a rousing rendition of the Snowball story to the assembled ceremonial crowd.

Some of Paul's critics still complain that he is resting too much of his reputation on this one idea. "Of all the people who need to prove themselves, I'd have thought Paul was the last," a researcher said to me one night, over the dregs of what had been a fine bottle of wine. "Whatever criterion you want to look at—he's a Harvard professor, he's fit, he's a member of the National Academy—apart from becoming Lord Hoffman, there's not much to go for, unless it's the big P. Posterity."

This isn't the whole story. Paul may love the attention that his work brings him, but his passion for the rocks themselves also runs deep. Working in the field, unearthing clues, and piecing them together into a picture that changes his understanding of the way the world works, that's when he feels fully alive. On the last day of a field season he often has tears in his eyes, though nobody has ever dared mention them. Paul goes to the rocks because he has to.

That combination of passion and aggrandizement has its inevitable effect. Paul sparks reactions in those around him. Some people are thrilled, others determined to cut him down to size. All are pulled into his story. This capacity to attract attention is crucial for really big scientific ideas. Theories like the Snowball often languish for decades without being properly probed. They need champions to drag them into the scientific limelight and expose them to scrutiny. They need people like Paul.

NOTES AND SUGGESTIONS
FOR FURTHER READING

Most of the material in this book comes from interviews with the researchers involved, from their students or former students, and from visits I made to their field sites. For some of the historical information, good general books or articles are available, and I've listed them in the notes that follow. But most of the research is so recent that published academic papers provide the only available accounts. I've included references to these papers, for the truly dedicated reader. Some of the research has not yet been published; in those cases information came directly from the researchers themselves.

ONE: FIRST FUMBLINGS

1. There's a wonderful account of the life and times of stromatolites in Ken McNamara's slim volume *Stromatolites,* 2nd edition (Perth: W.A. Museum, 1997); also see the vivid description of the creatures of Slimeworld in Jan Zalasiewicz and Kim Freedman, "The Dawn of Slime", *New Scientist,* 11 March 2000, 30.

2. Using 4.55 billion years for the origin of the Earth, 3.85 billion years for the earliest life, and .54 billion years for the Cambrian explosion.

3. John McPhee, *Basin and Range* (New York: Farrar, Straus & Giroux, 1981), 126.

4. Stephen Jay Gould, *Time's Arrow, Time's Cycle* (Cambridge, Mass.: Harvard University Press, 1998), 1–2.

5. Stephen Pyne is a true ice-lover, and his book *The Ice* (Seattle: University of Washington Press, 1998) pays detailed homage to all things frozen. Also see the wistful, wonderful *Arctic Dreams* by Barry Lopez (New York: Scribner's, 1986).

TWO: THE SHELTERING DESERT

1. Information about the expedition comes from George Whalley, *The Legend of John Hornby* (London: John Murray, 1962), and from Edgar Christian's diary, published by his parents after his death with the title *Unflinching: A Diary of Tragic Adventure* (London: John Murray, 1937).

2. Whalley, *The Legend of John Hornby*, 282.

3. R. F. Scott, *Scott's Last Expedition*, 5th edition, vol. 1 (London: Smith, Elder, 1914), 542.

4. Of the many, many words that have been written about why Scott's expedition went so wrong, Susan Solomon's book *The Coldest March* (New Haven and London: Yale University Press, 2001) is among the best. A world-renowned atmospheric scientist, Solomon analysed the meteorological data from Scott's trip and concluded that the adventurers encountered exceptionally bad weather. The best contemporary description of the events leading up to the tragedy is in Apsley Cherry-Garrard, *The Worst Journey in the World* (London: Picador, 1994).

5. See Simon Winchester, *The Map That Changed the World* (London: Viking, 2001).

6. P. H. Hoffman, "United Plates of America, the birth of a craton: Early Proterozoic assembly and growth of Laurentia", in *Annual Review of Earth and Planetary Sciences*, vol. 16 (1988), 543–603.

7. *The Ottawa Citizen*, 14 July 1989.

8. Paul borrowed this quote from John Kenneth Galbraith's novel *The Tenured Professor* (Boston: Houghton Mifflin, 1990). The character who made the remark was a former president of the University of California, Berkeley, who had been forced out after colliding with the governor of California, Ronald Reagan. Another character in the

book, hearing the remark, says, "He lost big because he won big. That's my idea of life." Paul told me he wished he had read the book earlier, so that he could have used the quote when he was actually leaving the Survey.

THREE: IN THE BEGINNING

1. Apsley Cherry-Garrard, *The Worst Journey in the World* (London: Picador, 1994).

2. Ibid., 369.

3. Here's an example of Brian's scrupulousness. When he was still doing research long after the age of retirement, he would carefully claim his own travel costs at the reduced rate for pensioners.

4. W. B. Harland, "The Cambridge Spitsbergen Expedition, 1949", *Geographical Journal* 118 (1952), 309–31.

5. Brian put forth this argument in a paper in the *Geological Magazine* 93, no. 4 (1956), 22.

6. See Edmund Blair Bolles, *The Ice Finders* (Washington D.C.: Counterpoint, 1999). Another good account of the development of ice age theories is John and Mary Gribbin's *Ice Age* (London: Penguin, 2001).

7. W. B. Harland, "Evidence of late Precambrian glaciation and its significance", in *Problems in Palaeoclimatology: Proceedings of the NATO Palaeoclimates Conference held at the University of Newcastle-upon-Tyne, January 7–12, 1963*, edited by A. E. M. Nairn (London: Interscience Publishers, 1964), 119.

8. J. C. Crowell, "Climate significance of sedimentary deposits containing dispersed megaclasts", in *Problems in Palaeoclimatology* (see above), 86.

9. Mike Hambrey, a student of Brian's who is now a geology professor at the University of Aberystwyth in Wales, performed the most detailed analysis of ice rocks in the late 1970s, and published the results in *Earth's Pre-Pleistocene Glacial Record*, edited by M. J. Hambrey and W. B. Harland (Cambridge, England: Cambridge University Press, 1984).

10. Alfred Wegener, *Annals of Meteorology* 4 (1951), 1–13.

11. H. W. Menard, *The Ocean of Truth: A Personal History of Global Tectonics* (Princeton, N.J.: Princeton University Press, 1986), 20–21.

12. "Continental Drift and Plate Tectonics: A Revolution in Science", in J. Bernard Cohen, *Revolution in Science* (Cambridge, Mass.: Harvard University Press, 1985), 446–66.

13. More on this story can be found in the rather uneasy account given by one of the researchers who survived through the winter at the station: Johannes Giorgi, *Mid-Ice: The Story of the Wegener Expedition to Greenland,* translated by F. H. Lyon (London: Kegan Paul, Trench, Trubner & Co., 1934); see also Martin Schwarzbach, *Alfred Wegener: The Father of Continental Drift* (Madison, Wisc.: Science Tech Publishers, 1986).

14. Mott T. Greene, "Alfred Wegener", *Social Research* 51, no. 3 (1984), 747.

15. In 1964 he laid out his main arguments in a paper titled "Critical evidence for a great infra-Cambrian glaciation", in *Geologische Rundschau* 54: 45–61; a more accessible version of the idea appears in a marvellous article that Brian wrote with M. J. S. Rudwick and published in *Scientific American* that same year, "The Great Infra-Cambrian Glaciation" (August 1964), 28.

16. W. B. Harland, "The Geology of Svalbard", *Geological Society Memoir* no. 17 (London: Geological Society, 1997).

FOUR: MAGNETIC MOMENTS

1. Joe later published this finding in "South-seeking magnetic bacteria", *Journal of Experimental Biology* 86 (1980), 345–47.

2. J. L. Gould, J. L. Kirschvink, and K. S. Deffeyes, "Bees have magnetic remanence", *Science* 201 (1978), 1026–28; C. Walcott, J. L. Gould, and J. L. Kirschvink, "Pigeons have magnets", *Science* 205 (1979), 1027–29.

3. J. L. Kirschvink, A. Kobayashi-Kirschvink, and B. J. Woodford, "Magnetite biomineralization in the human brain", *Proceedings of the National Academy of Sciences* 89 (1992), 7683–87.

4. The two researchers then incorporated additional information to make a somewhat more persuasive case, and Joe eventually recommended publication. The paper was published under the title "Low palaeolatitude of deposition for late Precambrian periglacial varvites in South Australia", *Earth and Planetary Science Letters* 79 (1986), 419–30. It did not, however, include the crucial field test that Joe went on to perform.

5. M. I. Budyko, "The effect of solar radiation variations on the climate of the Earth", *Tellus* 21 (1969), 611–19.

6. George E. Williams, "Precambrian tidal and glacial clastic deposits: Implications for Precambrian Earth-Moon dynamics and palaeoclimate", *Sedimentary Geology* 120 (1998), 55–74.

7. Dawn Sumner presented the results at the autumn meeting of the American Geophysical Union in 1987.

8. J. L. Kirschvink, "Late Proterozoic low-latitude glaciation: The Snowball Earth", section 2.3, in J. W. Schopf, C. Klein, and D. Des Maris, eds., *The Proterozoic Biosphere: A Multidisciplinary Study* (Cambridge, England: Cambridge University Press, 1992), 51–52.

FIVE: EUREKA

1. Paul went on to publish this paper, as well as the one with Dan Schrag, in *Science*. Paul F. Hoffman, Alan J. Kaufman, and Galen P. Halverson, "Comings and goings of global glaciations on a neoproterozoic carbonate platform in Namibia", *GSA Today* 8 (1998), 1–9.

2. Dan has made some marvellous discoveries about past climate in his coral work. See, for example, K. A. Hughen, D. P. Schrag, S. B. Jacobsen, and W. Hantoro, "El Niño during the last Interglacial recorded by fossil corals from Indonesia", *Geophysical Research Letters* 26 (1999), 3129–32. This tale is written up in more accessible form in "Weather warning", *New Scientist* 164 (9 October 1999), 36.

SIX: ON THE ROAD

1. P. F. Hoffman, A. J. Kaufman, G. P. Halverson, and D. P. Schrag, "A Neoproterozoic snowball Earth", *Science* 281 (1998), 1342–46. Paul

and Dan went on to write a more popular rendition of their ideas: "Snowball Earth", *Scientific American,* January 2000, 68–75.

2. Alfred Wegener, *The Origin of Continents and Oceans,* translated from the third German edition by J. G. A. Skerl (London: Methuen & Co., 1924), 5.

3. E. W. Berry comments on the Wegener hypothesis in *The Theory of Continental Drift: A Symposium,* edited by W. A. J. M. van Waterschoot van der Gracht (London: John Murray, 1928), 124.

4. B. Willis, *American Journal of Science* 242 (1944), 510–13.

5. For a more detailed discussion of these disagreements, see Naomi Oreskes's extremely thorough analysis, *The Rejection of Continental Drift* (Oxford, England: Oxford University Press, 1999).

6. From Mott T. Greene, "Alfred Wegener", *Social Research* 51, no. 3 (1984), 753.

7. Oreskes gives an excellent description of uniformitarianism in *The Rejection of Continental Drift* (see above). Also, Stephen Jay Gould has considered the principle in many fine essays. See, for example, his discussion of uniformitarianism and catastrophism, "Lyell's Pillars of Wisdom", in *The Lying Stones of Marrakech* (London: Vintage, 2000), 147–68; or the discussion in his *Time's Arrow, Time's Cycle* (Cambridge, Mass.: Harvard University Press, 1998).

8. Walter Alvarez wrote an entertaining book about this process, *T. rex and the Crater of Doom* (Princeton, N.J.: Princeton University Press, 1997).

9. W. M. Davis, "The value of outrageous geological hypotheses", *Science* 63 (1926), 464.

10. H. W. Menard, *The Ocean of Truth: A Personal History of Global Tectonics* (Princeton, N.J.: Princeton University Press, 1986).

11. Yes, Pippa is a man, and he has no idea why his parents burdened him with what appears to be a woman's name. In publications he goes by his first name, Galen, which he also finds baffling.

12. John Playfair, *Illustrations of the Huttonian Theory of the Earth* (Edinburgh: William Creech, 1802).

13. Gould, *Time's Arrow, Time's Cycle,* 64.

14. ———, "James Hutton's Theory of the Earth", in *Time's Arrow, Time's Cycle,* 61–98.

SEVEN: DOWN UNDER

1. Linda Sohl, Nicholas Christie-Blick, and Dennis Kent, "Paleomagnetic polarity reversals in Marinoan glacial deposits of Australia", *GSA Bulletin* 111 (1999), 1120–39.
2. George Williams has described his idea about the tilting of the Earth in a series of academic papers. The broadest and best for nonspecialists is probably his chapter entitled "The enigmatic Late Proterozoic glacial climate: An Australian perspective", in *Earth's Glacial Record* (Cambridge, England: Cambridge University Press, 1994), 146–64.
3. *The Australian Geologist* 117 (31 December 2000), 21.
4. After some heated e-mail exchanges with the editor of *The Australian Geologist,* Paul abandoned his attempt to publish a ten-page rebuttal of George Williams's arguments in the magazine. Instead he posted his rebuttal on his website, http://www.eps.harvard.edu/people/faculty/hoffman/TAG.html
5. There is even some evidence that the presence of the moon *prevented* the Earth's tilt from further fluctuations. See, for example, J. Laskar, F. Joutel, and P. Robutel, "Stabilization of the Earth's obliquity by the Moon", *Nature* 361 (1993), 615–17. Recently one group of researchers did try to come up with a possible mechanism for righting the Earth: See Darren Williams, James Kasting, and Lawrence Frakes, "Low-latitude glaciation and rapid changes in the Earth's obliquity explained by obliquity-oblateness feedback", *Nature* 396 (1998), 453. But the authors say that their paper serves to demonstrate how difficult the task would be.
6. Jim Walker's paper is now in press at the *Proceedings of the National Academy of Sciences.*

NOTES AND SUGGESTIONS FOR FURTHER READING

EIGHT: SNOWBRAWLS

1. Once I was interviewing Martin Kennedy in an Italian restaurant, using a minidisk recorder and a microphone. When Martin left for the bathroom, the proprietor sidled up to me and asked, "Who is he? Why are you interviewing him? Is he someone famous?" When Martin returned, I related this story and he grinned. "Surely he's heard of the *Kennedys*," he said in a stage whisper. Then, "Did he ask what kind of book you're writing? You did tell him, didn't you, that it's a thriller?"

2. For more on the remarkable properties of methane hydrates, see Erwin Suess, Gerhard Bohrmann, Jens Greinert, and Erwin Lausch, "Flammable ice", *Scientific American*, November 1999, 76–83. There's also a vivid essay by Nicola Jones: "Fire and Ice", *Chemistry and Industry* 26 (June 2000), 398–99.

3. Martin Kennedy, Nicholas Christie-Blick, and Linda Sohl, "Are Proterozoic cap carbonates and isotopic excursions a record of gas hydrate destabilization following Earth's coldest intervals?" *Geology* 29, no. 5 (2001), 443–46.

4. Martin Kennedy, Nicholas Christie-Blick, and Anthony Prave, "Carbon isotopic composition of Neoproterozoic glacial carbonates as a test of paleoceanographic models for snowball Earth phenomena", *Geology* 29, no. 12 (2001), 1135–38.

5. "The Aftermath of a Snowball Earth", by John Higgins and Daniel Schrag, submitted to the electronic journal *Geochemistry, Geophysics, Geosystems*.

6. Though Huxley's reply is widely quoted, the precise wording varies between versions, and sadly no verbatim account of the debate exists. See, for example, *The Columbia World of Quotations*, edited by R. Andrews, M. Biggs, and M. Seidel (New York: Columbia University Press, 1996).

7. This quote comes from the marvellous *Ice Palaces* by Fred Anderes and Ann Agranoff (New York: Abbeville Press, 1983). Sadly, the book is out of print, but it is well worth hunting around for a used copy.

8. Douglas Mawson, "The Home of the Blizzard", (New York: St.

Martin's Press, 1998), xvii. One of the best books ever written about Antarctic exploration, and yet little known outside Australia, this is a must for anyone who cares about ice.

9. See Philip Ball's vivid descriptions of this in *Life's Matrix: A Biography of Water* (Berkeley: University of California Press, 2001).

10. For example, the paper by William Hyde, Thomas Crowley, Steven Baum and Richard Peltier, "Neoproterozoic 'snowball Earth' simulations with a coupled climate/ice-sheet model", *Nature* 405 (2000), 425–29; also Bruce Runnegar's commentary, "Loophole for snowball Earth", on page 403 of the same issue; and Mark Chandler and Linda Sohl, "Climate forcings and the initiation of low-latitude ice sheets during the Neoproterozoic Varanger glacial interval", *Journal of Geophysical Research* 105 (2000), 20, 737–20, 756.

11. Doug and Dan are now writing up this work for publication.

NINE: CREATION

1. The dating is controversial, with a few Snowball deposits dated as early as 575 million years.

2. A wonderful book! Stephen Jay Gould, *Wonderful Life* (New York: Vintage, 2000).

3. Attempts to explain the *form* of the Cambrian explosion have been many and various. The fact that such an explosion ever occurred, though, can be traced back to the previous leap into multicellularity. For more on this, see Carl Zimmer, *Evolution: The Triumph of an Idea* (London: William Heinemann, 2002); and Bill Schopf, *Cradle of Life* (Princeton, N.J.: Princeton University Press, 1999).

4. A few Ediacarans had been found before Sprigg's discovery, for example from Charnwood Forest, in England. But they were not grouped together or given this collective name until the major finds in South Australia.

5. James Gehling, "Microbial Mats in terminal Proterozoic siliciclastics: Ediacaran death masks", *Palaios* 14 (1999), 40–57.

6. Researcher Mark McMenamin has written a book titled *The Garden of Ediacara* (New York: Columbia University Press, 1998), in which

he argues that the Ediacarans were a failed experiment, which then became extinct. (Be warned, though, this book was received with little enthusiasm by the rest of the Ediacaran community, and one researcher said in a review that it "falls as flat as a week-old *Dickinsonia* roadkill".) For more on this debate, see also Bennett Daviss, "Cast out of Eden", *New Scientist* 158 (16 May 1998), 26; and Richard Monastersky, "Life grows up", *National Geographic*, April 1998, 100–15. In an academic treatise, Jim Gehling sets out his arguments that some at least of the Ediacarans evolved into more familiar animals: "The case for Ediacaran fossil roots to the metazoan tree", *Geological Society of India Memoir No. 20* (1991), 181–224.

7. The researcher is Ben Waggoner, now at the University of Central Arkansas, and the fossil he is so proud of is called *Yorgia waggoneri*. He says he finds the fact that nobody knows the exact nature of this creature "somehow deeply appropriate and satisfying".

8. These finds are all extremely recent. Misha is now preparing his descriptions and conclusions for publication.

9. A. P. Benus, *Bulletin of New York State Museum* 463 (1988), 8.

10. There is no sign of any animal trails among the Ediacarans at Mistaken Point. Since these creatures are much older than the ones found in South Australia or the White Sea, they are presumably at an earlier state of evolution. Many researchers believe that their large size and the intricate fronds and spindles mean that these, too, are complex, differentiated creatures.

11. S. A. Bowring et al., "Geochronological constraints on the duration of the Neoproterozoic-Cambrian transition", *Geological Society of America,* 1998 annual meeting, Abstract A147; S. A. Bowring and D. H. Erwin, "Progress in Calibrating the Tree of Life and Metazoan Phylogeny", *Eos Trans. AGU* 81 (48), Fall Meeting Supplement, Abstract B62A–04, 2000.

12. H. J. Hofmann, G. M. Narbonne, and J. D. Aitken, "Ediacaran Remains from Intertillite Beds in Northwestern Canada", *Geology* 18 (1990), 1199–1202.

13. A. Seilacher, P. K. Bose, and F. Pflüger, "Triploblastic animals more than 1 billion years ago: Trace fossil evidence from India", *Science* 282 (1998), 80.

14. B. Rasmussen, S. Bengston, I. Fletcher, and N. McNaughton, "Discoidal impressions and trace-like fossils more than 1200 million years old", *Science* 296 (2002), 1112–15.

15. N. J. Butterfield, *"Bangiomorpha pubescens:* Implications for the evolution of sex, multicellularity, and the Mesoproterozoic/Neoproterozoic radiation of eukaryotes", *Paleobiology* 26 (2000), 386–404.

16. There's a good, though somewhat technical, overview of this technique in Andrew Smith and Kevin Peterson, "Dating the time of origin of major clades: Molecular clocks and the fossil record", *Annual Reviews of Earth and Planetary Science* 30 (2002), 65–89.

17. Lee Smolin, "Art, science and democracy", written for a catalogue of an exhibit of sculpture by Elizabeth Turk at the Santa Barbara Contemporary Arts Forum, 24 February–14 April 2001.

18. Kevin is currently preparing his paper for publication.

TEN: EVER AGAIN

1. Joe published these results in D. A. Evans, N. J. Beukes, and J. L. Kirschvink, "Low-latitude glaciation in the Palaeoproterozoic era", *Nature* 386 (1997), 262–65. His primitive Snowball didn't seem to trigger the same dramatic evolutionary changes that accompanied the later one. It may simply have been too early. For any evolutionary leap, living things don't just need the environmental opportunity, they also need the genetic wherewithal. Genetic material changes through time and chance, and the creatures living during this early Snowball may not yet have had long enough to string together the genes they'd eventually need.

2. Dan Schrag also points out that rainfall in the tropics is much more intense than at higher latitudes. The more intense the rainfall, the more effective rocks are at trapping carbon dioxide. If all the world's continents were near the equator, their collective ability to suck up carbon dioxide would go into overdrive; both advancing ice

and continental rocks would be working in tandem to cool the Earth. He put all these arguments in a paper in which he also puts forward an intriguing idea for what specifically may have triggered the cooling that led to one of Paul's suite of Snowballs. G. P. Halverson, P. F. Hoffman, D. P. Schrag, and A. J. Kaufman, "A major perturbation of the carbon cycle before the Ghaun glaciation in Namibia: prelude to snowball Earth", *Geochemistry, Geophysics, Geosystems,* 27 June 2002.

3. David Evans, "True polar wander, a supercontinental legacy", *Earth and Planetary Science Letters* 157 (1998), 1–8; and D. A. Evans, "True polar wander and supercontinents", *Tectonophysics,* 2002, in press. A more accessible description of this idea is Robert Irion, "Slip-sliding away", *New Scientist* 18 (August 2001), 34.

4. For a good general description of the motions of continents, see David M. Harland *The Earth in Context* (Chichester, England: Springer-Praxis, 2001). There is a fuller and more technical description in Donald Turcotte and Gerald Schubert *Geodynamics,* 2nd edition (Cambridge, England: Cambridge University Press, 2002).

5. Chris Scotese's construction of the possible new supercontinent, which he calls "Pangaea Ultima", is part of his Paleomap Project. More information about this is on his website, www.scotese.com.

EPILOGUE

1. *The Biggest Snowball Ever.* Copyright © 1988 by John Rogan. Reproduced by permission of the publisher, Candlewick Press, Inc., Cambridge, MA, on behalf of Walker Books Ltd., London.

INDEX

INDEX

A NOTE ON THE AUTHOR

Gabrielle Walker has travelled in search of science stories to all seven continents - including a stint at the South Pole. She has climbed trees in the Amazon rainforest, used a geological hammer to pull fresh lava from a volcano in Hawaii and dodged icebergs while sailing across Drake's passage and around Cape Horn. She has a PhD in Natural Science from Cambridge University and has been Editor at *Nature* and Features Editor at *New Scientist* for whom she now acts as consultant. An award-winning writer of more than sixty pieces for *New Scientist*, she has also written for *The Economist*, the *Independent*, the *Guardian* and the *Daily Telegraph*, and is a frequent studio guest and presenter for the BBC. She lives in London. This is her first book.